# Global Changes and Natural Disaster Management: Geo-information Technologies

Saied Pirasteh · Jonathan Li
Editors

# Global Changes and Natural Disaster Management: Geo-information Technologies

Springer

*Editors*
Saied Pirasteh
Institute of Disaster Management (WIDM)
Mississauga, ON
Canada

and

Department of Geography & Environmental Management
University of Waterloo
Waterloo, ON
Canada

Jonathan Li
Department of Geography & Environmental Management
University of Waterloo
Waterloo, ON
Canada

Every effort has been made to contact the copyright holders of the figures and tables which have been reproduced from other sources. Anyone who has not been properly credited is requested to contact the publishers, so that due acknowledgment may be made in subsequent editions.

ISBN 978-3-319-51843-5 ISBN 978-3-319-51844-2 (eBook)
DOI 10.1007/978-3-319-51844-2

Library of Congress Control Number: 2016963191



Printed on acid-free paper

This Springer imprint is published by Springer Nature
The registered company is Springer International Publishing AG
The registered company address is: Gewerbestrasse 11, 6330 Cham, Switzerland

*Dedicated to Prof. Ali Rashid Al Noaimi, the vice chancellor of the United Arab Emirates University; Prof. Dr. Saif Al-Qaydi, dean of College of Humanities and Social Sciences; Dr. Khaula Alkaabi, chair of Geography and Urban Planning Department; Dr. Mohamed Yagoub, former chair of Geography and Urban Planning Department; and Dr. Khalid Hussein, Mr. Abdallah Al-Bizreh, students, and people at the United Arab Emirates.*

# Preface

During the last few decades, we have seen disasters not only stem from natural and technological sources but human activities have affected societies with increasing frequency and global change as well. This resulted in enormous socioeconomic losses and damaged the lives in both developing and developed countries.

State-of-the-art technologies have implemented designing various tools and applications for saving the earth with informed solutions. Geo-informatics technologies and earth observation satellites have already demonstrated a strong flexibility in providing data, analyzing data, and modeling for a broad range of applications. The Geo-informatics for Natural Disaster Management (GiT4NDM) and Earth Observation Global Changes (EOGC) events have been struggling in establishing a platform for scientists, researchers, practitioners, etc. The goal was to make a sustainable development by organizing events around the world. This began from 2007, and it is still ongoing (www.widm.ca). Thus, the GiT4NDM events known for their "Informed Solution and Diversity" concern not only earth observation systems but also hazards and disasters. The concerns are in global environmental water issues, climate changes, hazards, disaster management, geospatial analysis and modeling, digital image processing, mapping, and management of natural resources, which have also been complex tasks.

State-of-the-art technologies' sectors can represent a significant portion of the socioeconomy of every industrialized nation in the world. Thus, this indicated how technologies are important for every country. However, technology has also become ubiquitous in modern society and nowadays we talk about "High Technology." Many technologies have been showing tremendous promise since the beginning of the twenty-first century. These include earth observation systems, laser scanning systems, Internet advancements, computer devices, and software advancements related to hazards, disaster management, energy, and the environment. The GiT4NDM motivates researchers to explore and find the intersection of these technologies that may indeed enable the most promising opportunities.

In this context, the editors of this volume had a long-time paper selection process extracted from 169 papers submitted to the 7th GiT4NDM-5th EOGC 2015 (December 8–10, 2015, UAE University, UAE). After the peer-reviewing process by the editors and assistant editors, the authors of a total of 51 full papers were invited to submit the revised version of the manuscripts. However, in total 16 revised version manuscripts were again evaluated for publication in the current volume.

This book contains six parts: (1) Land Use and Land Cover Change, (2) Agriculture Monitoring, (3) Smart City, (4) Climate Change, (5) Risk Assessment, (6) Disaster Management.

Visiting Al-Ain schools' students to the 7th GiT4NDM-5th EOGC 2015 during the conference

Exhibition at the 7th GiT4NDM-5th EOGC 2015

A special thanks to Leila Farzinpur and Sobhan Pirasteh who support me in every stage of my career. Finally, I thank Prof. Dr. Saif Al-Qaydi (dean of College of Humanities and Social Sciences); Dr. Khaula Alkaabi (chair of Geography and Urban Planning Department); Dr. Mohamed Yagoub (former chair of Geography and Urban Planning Department); and Dr. Khalid Hussein and Mr. Abdallah Al-Bizreh from the Geography and Urban Planning Department for their full support in making the 7th GiT4NDM-5th EOGC 2015 and publishing this book very successful. I am also thankful to Yifei Chen and Yuenan Li (Master's degree students of University of Waterloo) for assisting me during the editing of the chapters.

Mississauga/Waterloo, Canada　　　　Saied Pirasteh

# Contents

# Contributors

**Adam M. Abbas** Department of Geography, Federal University Kashere, Gombe-State, Nigeria

**Rifaat Abdalla** Department of Hydrographic Surveying, Faculty of Maritime Studies, King Abdulaziz University, Jeddah, Kingdom of Saudi Arabia

**U.A. Abubakar** Department of Political Science, Federal University Kashere, Gombe State, Nigeria

**A. Ahmed** Department of Political Science, Federal University Kashere, Gombe State, Nigeria

**Yusuf A. Aina** Department of Geomatics Engineering Technology, Yanbu Industrial College, Yanbu, Saudi Arabia

**Salim Mubarak Al Hatrushi** Department of Geography College of Arts and Social Sciences, Sultan Qaboos University, Muscat, Oman

**Rami Al-Ruzouq** Civil and Environmental Engineering Department, University of Sharjah, Sharjah, United Arab Emirates

**Bashir Almurad** United Arab Emirates University, Al-Ain, UAE

**Ali Asgary** Disaster and Emergency Management, School of Administrative Studies, York University, Toronto, Canada

**Omosalameh Babaee** Department of Geography, University of Payame Noor, Tehran, Tehran Province, Iran

**Abdul-Lateef Balogun** Department of Geomatics Engineering Technology, Yanbu Industrial College, Yanbu, Saudi Arabia; Geomatics Research Group, Department of Civil And Environmental Engineering, Universiti Teknologi PETRONAS, Perak, Malaysia

**A. Bannari** Geoinformatics Department, College of Graduate Studies, Arabian Gulf University, Manama, Kingdom of Bahrain

**A.L. Costa** Masdar Institute of Science and Technology, Abu Dhabi, UAE

**Weihong Cui** School of Remote Sensing and Information Engineering, Wuhan University, Wuhan, China; Collaborative Innovation Center for Geospatial Technology, Wuhan, China; Mobile Mapping Lab, University of Waterloo, Waterloo, Canada

**A. El-Battay** Geoinformatics Department, College of Graduate Studies, Arabian Gulf University, Manama, Kingdom of Bahrain

**Chenyi Feng** Xi'an University of Science and Technology, Xi'an, China

**A. Ghadeer** Geoinformatics Department, College of Graduate Studies, Arabian Gulf University, Manama, Kingdom of Bahrain

**Khaled Hamad** Civil and Environmental Engineering Department, University of Sharjah, Sharjah, United Arab Emirates

**N.A. Hameed** Geoinformatics Department, College of Graduate Studies, Arabian Gulf University, Manama, Kingdom of Bahrain

**H. Hussein** Department of Electrical Engineering, United Arab Emirates University, Abu Dhabi, UAE

**M. Hussein** Department of Electrical Engineering, United Arab Emirates University, Abu Dhabi, UAE

**S. Issa** Geology Department, United Arab Emirates University, Al-Ain, United Arab Emirates

**Daryoush Kari** Communication & Culture Studies, York University, Toronto, Canada

**Jonathan Li** Mobile Mapping Lab, University of Waterloo, Waterloo, Canada

**W. Li** Massachusetts Institute of Technology, Cambridge, USA

**Yuenan Li** Department of Geography and Environmental Management, University of Waterloo, Waterloo, Canada

**Kai Liu** School of Geography and Planning, Sun Yat-sen University, Guangzhou, China

**Yang Liu** Department of Geography, University of Cincinnati, Cincinnati, USA

**Peyman Molaali** Department of GIS and Cadaster, Khuzestan Water and Power Authority, Ahvaz, Khuzestan Province, Iran

**Aman Mwafy** United Arab Emirates University, Al-Ain, UAE; Faculty of Engineering, Zagazig University, Zagazig, Egypt

**M. Opolot** Masdar Institute of Science and Technology, Abu Dhabi, UAE

**Irshad M. Parvez** Department of Geomatics Engineering Technology, Yanbu Industrial College, Yanbu, Saudi Arabia

**M. Rouai** Department of Earth Sciences, Faculty of Sciences, University Moulay Ismail, Meknes, Morocco

**R. Saeed** RS/GIS Postgrad, United Arab Emirates University, Al-Ain, United Arab Emirates

**N. Saleous** Geography Department, United Arab Emirates University, Al-Ain, United Arab Emirates

**Abdallah Shanableh** Civil and Environmental Engineering Department, University of Sharjah, Sharjah, United Arab Emirates

**R.L. Sousa** Masdar Institute of Science and Technology, Abu Dhabi, UAE

**Guofeng Wang** China Highway Engineering Consulting Corporation, Beijing, China

**Xiukang Wang** College of Life Science, Yan'an University, Yan'an, Shaanxi, People's Republic of China; Northwest Arid Regions of Water Conservancy Engineering State Key Laboratory, Xi'an University of Technology, Xi'an, Shaanxi, People's Republic of China

**Yingying Xing** College of Life Science, Yan'an University, Yan'an, Shaanxi, People's Republic of China

**Yiwei Zheng** Xi'an University of Science and Technology, Xi'an, China

**Yuanhui Zhu** School of Geography and Planning, Sun Yat-sen University, Guangzhou, China

# Part I
# Land Use and Land Cover Change

# Examining the Effect of Land Use on the Spatiotemporal Dynamics of Urban Temperature in an Industrial City: A Landsat Imagery Analysis

**Yusuf A. Aina, Irshad M. Parvez and Abdul-Lateef Balogun**

**Abstract** The urban heat island (UHI) effect is a human-induced phenomenon that results in higher temperatures in urbanized areas as compared to their surroundings. Yanbu industrial city is one of the highly industrialized cities in Saudi Arabia with petrochemical, energy-intensive activities, and a growing population. So, it is imperative to study the effect of the industrial activities on the changes in temperatures especially in an arid area where such studies are limited. The objective of this study is to decipher the spatiotemporal variations in the temperatures of different land use/land cover types in Yanbu industrial city. Thermal bands of Landsat satellite images were used in the study. Multi-temporal images that spread over a period of 25 years between 1990 and 2015 were used to extract the land surface temperatures of the study area. Different land use/land cover categories in the study area were identified from IKONOS false color composite image (2012), and a correlation was sought out, between vegetationand surface temperature, using a vegetation index (NDVI). Normalized difference build index (NDBI) was derived in order to analyze the relationship between surface temperature and imperviousness. The results depict the thermal regimes of the area in different LULC types and their variations over time. The NDVI has an inverse relation with temperature whereas the NDBI has a positive relationship with the surface temperature, only in the residential and conservation areas. Urban heat islands are detected in the industrial area with 5–7° higher surface temperature, and urban cool islands are detected in the residential area with 3–5° lower surface temperature.

Y.A. Aina (✉) · I.M. Parvez · A.-L. Balogun
Department of Geomatics Engineering Technology,
Yanbu Industrial College, Yanbu, Saudi Arabia
e-mail: ainay@rcyci.edu.sa

I.M. Parvez
e-mail: miri@rcyci.edu.sa

A.-L. Balogun
e-mail: baloguna@rcyci.edu.sa

A.-L. Balogun
Geomatics Research Group, Department of Civil And Environmental
Engineering, Universiti Teknologi PETRONAS, Perak, Malaysia

 
S. Pirasteh and J. Li (eds.), *Global Changes and Natural Disaster Management: Geo-information Technologies*, DOI 10.1007/978-3-319-51844-2_1

**Keywords** Landsat · Land surface temperature · NDVI · NDBI · Urban heat island · Urban cool island

## 1 Introduction

One of the problems of rapid urbanization is the urban heat island (UHI). This is a phenomenon whereby there are notable differences between rural and urban temperature regimes. The phenomenon is attributable to the environmental impacts of anthropogenic activities (Deng et al. 2014). Anthropogenic activities in urban areashave made them areas of complex environmental interactions and this complexity has been affecting the exchange of energy through changes in the physical environment (Yusuf et al. 2014). It has become imperative to monitor the dynamics of UHI in urban environments since changes in heat budgets have impacts on global climate change (Deng et al. 2014). Due to the lack of adequate coverage of urban centers by weather stations and the capabilities of remote sensingtechnologies, satellite data have been adopted in studying the relationship between land use change dynamics and heat budget in urban areas (Buyantuyev and Wu 2010; Chen et al. 2006). Satellite data can be used to assess the temperature variations in time and space in an urban environment with the view to mitigating the effects of human activities. Quite a number of studies (Yusuf et al. 2014; Buyantuyev and Wu 2010; Chen et al. 2006; Schwarz et al. 2011) have examined the spatiotemporal variations in UHI at different urban environments. Some studies have even highlighted the health implications of high surface temperatures in urban areas (Jenerette et al. 2015; Vanos et al. 2016). In Saudi Arabia, there are few studies on UHI (Lim et al. 2007, 2012; Alghamdi and Moore 2015) especially studies that analyze the dynamics of urban heat island in different land covertypes. This study examines the variations in UHI in an industrial city of Yanbu by using Landsatimages from 1991 to 2015. It also assesses the relationship between land surface temperature, Normalized difference vegetation index (NDVI), and Normalized difference build index (NDBI) in the desert environment.

## 2 Study Area

The study was conducted in the coastal Yanbu industrial city (Fig. 1). The city is located at the North of the second largest city, Jeddah, in Saudi Arabia. Established in the early 1970s, Yanbu's strategic location makes it attractive to many industrial organizations. It is one of the major ports in Saudi Arabia and a vital shipping terminal for transporting petroleum products. It is also a rapidly expanding city covering a cumulative area of 606 $km^2$, including existing (185 $km^2$) and expansion areas (420 $km^2$). The city has a refining capacity of 1.1 million barrel/day and accommodates 212 industries and 747 commercial establishments respectively

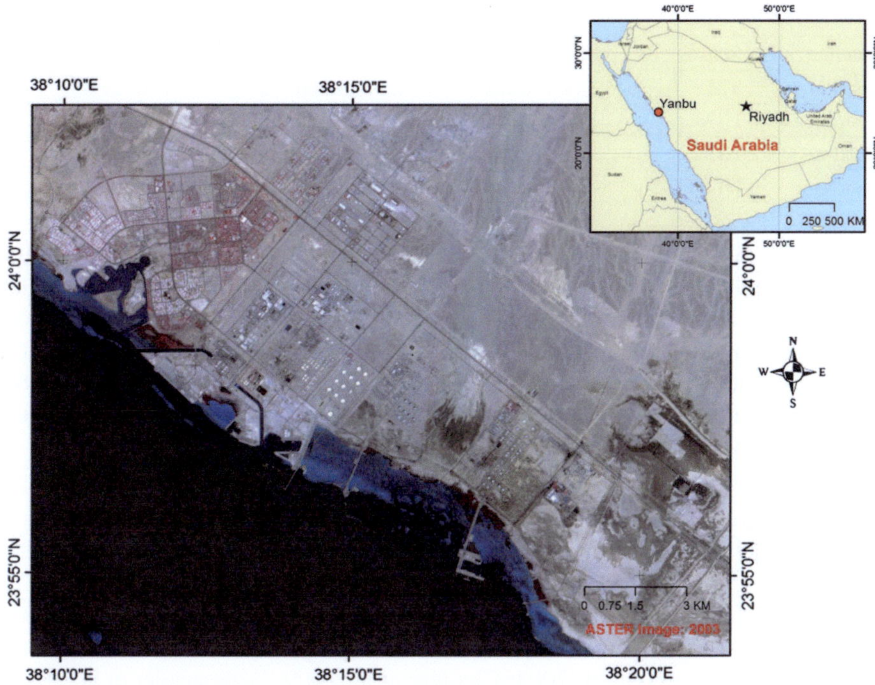


**Fig. 1** Location of the study area

(RCY 2015). The budget of the city's commission in 2012 was about 2.4 billion Saudi Riyals (Aina and Aleem 2014). In contrast to many densely populated coastal cities, Yanbu is moderately populated employing 55,030 people (RCY 2015). The current population of the city is 117,936 which is still lower than the planned capacity (285,000) of the city (RCY 2015). Its massive industrial activities make it an ideal choice for this study's analysis.

## 3 Methodology

In this study, Landsat data encompassing the period between 1991 and 2015, collected using TM, ETM, OLI, and TIRS sensors, had been used in the analysis. The data was downloaded from the USGS Earth Explorer data portal. The Landsat series of sensors are radiometers that record electromagnetic energy in the visible, infrared, and thermal portions of the electromagnetic spectrum. The visible portion of the spectrum primarily comprise of the solar energy reflected from the different objects on the Earth's surface along with the signatures from the atmosphere, whereas the thermal sensors onboard measure the emitted radiations (Schowengerdt 2007) of the objects that hinged on the thermal characteristic of the emitting body and encompassed their heat capacity, thermal diffusivity, and thermal inertia.

**Table 1** Source data

| Satellite | Sensor | Bands | Spatial resolution (m) | Acquisition date |
|---|---|---|---|---|
| Landsat-5 | TM | 3, 4, 5 and 6 | 30, 30, 30 and 120 | Feb, 1991 |
| Landsat-7 | ETM | 3, 4, 5 and 6 | 30, 30, 30 and 60 | Dec, 2000 |
| Landsat-7 | ETM | 3, 4, 5 and 6 | 30, 30, 30 and 60 | Jan, 2007 |
| Landsat-8 | OLI/TIRS | 4, 5, 6 and 10 | 30, 30, 30 and 100 | Jan, 2015 |

Visible, near infrared, and thermal bands of the above-cited sensors, were used to derive biophysical parameters like NDVI, NDBI, and land surface emissivity which in turn served as inputs to derive the land surface temperature of the study area from the multi-temporal Landsatsatellite images. The sensors, the bands, and the acquisition date of the data are highlighted in Table 1. All the images were acquired around winter season when the temperature of the area is relatively low, and vegetation cover is relatively high.

## *3.1 Retrieval of Land Surface Temperature (LST)*

The methodology followed in this study to derive LST from the selected Landsat/ TM, ETM+, and TIRS data consist of the following steps (Jimenez-Munoz and Sobrino 2003; Sobrino et al. 2004; Chander et al. 2009; Walawender et al. 2014):

Conversion of pixel values (DN values) of the Landsat thermal bands of Level-1 Product Generation System (LPGS) to at-sensor spectral radiance (Ls) in W $m^{-2}$ $sr^{-1}$ $\mu m^{-1}$ according to Eq. (1):

$$\mathrm{Ls} = \mathrm{gain} * \mathrm{DN} + \mathrm{bias}, \tag{1}$$

where DN represents the digital number (the pixel value) in level 1 (LPGS); and gain (the slope of the radiance) and bias (the intercept of the radiance) are the DN conversion coefficients as given in Table 2 (Chander et al. 2009).

The spectral radiance at-sensor Ls calculated in the step above was converted to at-sensor brightness temperature Ts using inverted Planck's Law using Eq. (2) (Irish 2003) (Table 3).

**Table 2** DN conversion coefficients

| Sensor | Band | Gain | Bias |
|---|---|---|---|
| TM | 6L | 0.055158 | 1.2378 |
| ETM+ | 6L | 0.0 | 17.04 |
| TIRS | 10 | 0.000334 | 0.10000 |

**Table 3** Thermal band calibration constants

| Sensor | $K_1$ (W $m^{-2}$ $sr^{-1}$ $\mu m^{-1)}$ | $K_2$ (K) |
|---|---|---|
| TM | 607.76 | 1260.56 |
| ETM+ | 666.09 | 1282.71 |
| TIRS | 774.8853 | 1321.0789 |

$$\mathrm{Ts} = K_2 / \ln(K_1/\mathrm{Ls} + 1) \quad (2)$$

where $K_1$ and $K_2$ are specific thermal band calibration constants for the respective sensors (Chander et al. 2009). The conversion of the at-sensor spectral radiance to at-sensor brightness temperature is based on the assumption that the surface of the earth is a perfect blackbody with an emissivity value of 1 (Irish 2003; Li et al. 2009, 2011). The land surface temperature was calculated using the single-channel (SC) algorithm, which relates at-sensor brightness temperature to the land surface temperatureincorporating land surface emissivity and various atmospheric parameters.

Single-channel (SC) algorithm estimates LST values by incorporating land surface emissivity along with various atmospheric functions using Eq. (3) (Jimenez-Munoz and Sobrino 2003; Sobrino et al. 2004; Jimenez-Munoz et al. 2009; Cristobal et al. 2009).

$$\mathrm{LST} = \gamma[1/\varepsilon(\psi 1\mathrm{Ls} + \psi 2) + \psi 3] + \delta \quad (3)$$

where $\varepsilon$ is the land surface emissivity (LSE); $\gamma$ and $\delta$ are constraints that depend on Plank's function; and $\psi 1$, $\psi 2$ and $\psi 3$ are the atmospheric functions.

The constraints $\gamma$ and $\delta$ were estimated using Eqs. (4 and 5):

$$\gamma = \{c2\mathrm{Ls}/T2s[\lambda 4\mathrm{Ls}/c1 + 1/\lambda]\} - 1 \quad (4)$$

$$\delta = -\gamma \cdot \mathrm{Ls} + \mathrm{Ts} \quad (5)$$

where $c1$ and $c2$ are the Planck's radiation constants ($c1 = 1.19104 \times 10^8$ W μm$^4$ m$^{-2}$ sr$^{-1}$; $c2 = 1.43877 \times 10^4$ μm K); and $\lambda$ is the effective wavelength for each band used ($\lambda = 11.43$, 11.33 and 10.9 μm for Landsat-5/TM, Landsat-7 ETM+, and Landsat-8 TIRS respectively).

The single-channel (SC) algorithm includes various parameters to decipher the influence of atmosphere on the emission properties of various land surface features. Before deriving LST, it is imperative to identify the atmospheric influences on the radiations recorded by the sensor. Atmospheric corrections are to be calculated by incorporating site-specific parameters (atmospheric transmissivity $\tau$, upwelling atmospheric radiance $L\uparrow$, and downwelling atmospheric radiance $L\downarrow$) to quantize its influence. These parameters were calculated using NASA's web-based tool known as Atmospheric Correction Parameter Calculator (ACPC) (Barsi et al. 2003, 2005). The ACPC is available online at http://atmcorr.gsfc.nasa.gov. This tool uses the National Center for Environmental Prediction (NCEP)-modeled atmospheric global profiles for a particular date, time, and location as input. Using MODTRAN software and a set of integration algorithms, the site-specific atmospheric transmission, upwelling and downwelling radiances were derived. The ACPC web tool provides an option for incorporating the local surface conditions such as temperature, altitude, pressure, and relative humidity. In the absence of these surface parameters, the model predicted surface conditions were incorporated. In the current study, the atmospheric data provided by the Jeddah Regional Climate Center (available at http://jrcc.sa) were used to compute the parameters. The site-specific atmospheric parameters ($\tau$, $L\uparrow$ and $L\downarrow$) calculated using ACPC were then used to calculate atmospheric functions ($\psi1$, $\psi2$, and $\psi3$) which were necessary input variables for the SC method in Eq. (3). The atmospheric functions were defined as given in Eqs. (6,7, and 8) (Jenerette et al. 2015; Vanos et al. 2016; Lim et al. 2012):

$$\psi1 = 1/\tau \quad (6)$$

$$\psi2 = -L\downarrow - L\uparrow/\tau \quad (7)$$

$$\psi3 = L\downarrow \quad (8)$$

## Estimation of Land Surface Emissivity

To estimate the land surface temperature using single-channel algorithm, the land surface emissivity of the area has to be computed apriori using Eq. (9). In the present study, NDVI threshold as calculated by using Eq. (10) had been employed which offers quite a simplistic approach to calculating LSE (Sobrino and Raissouni 2000; Valor and Caselles 1996; Van De Griend and Owe 1993). The NDVI Threshold Method which relies on identifying the pure end members as vegeta-tionand bare soil, derives the emissivity values of land cover features from the NDVI values considering different scenarios: (a) If NDVI is less than 0.2, the pixel is attributed to bare soil, and red band reflectivity values are used to get the emissivity. (b) If NDVI is greater than 0.5, pixels falling in this class are treated as fully vegetated, and then a constant value for the emissivity is assumed, typically of 0.99. (c) If NDVI is greater than 0.2 and less than 0.5, this class corresponds to mixed pixels wherein they have a varying mixture of bare soil and vegetation and the emissivity is calculated according to Eq. (9):

$$\varepsilon = \varepsilon v Pv + \varepsilon s(1 + Pv) + d\varepsilon \quad (9)$$

where $\varepsilon v$ is the emissivity of vegetation and $\varepsilon s$ is the soil emissivity, Pv is the vegetationproportion obtained using Eq. (11) (Carlson and Ripley 1997), $d\varepsilon$ is the term for geometrical distribution effect and internal reflections. This method needs soil and vegetation emissivity. To apply this methodology, values of soil and vegetation emissivity are needed which were calculated to be 0.99 and 0.97 respectively.

$$\mathrm{NDVI} = \mathrm{NIR} - \mathrm{RED}/\mathrm{NIR} + \mathrm{RED} \quad (10)$$

In this present study, the general assumptions of the NDVI-based land surface emissivity method were simplified. Values of NDVIS = 0.2 (for soil) and NDVIV = 0.5 (for vegetation) proposed by Sobrino and Raissouni (2000) for global conditions were adjusted to suit the study area conditions. Using the NDVIS and NDVIV, fractional vegetation cover was derived from NDVI using Eq. (11) (Carlson and Ripley 1997).

$$PV = (NDVI - NDVIS/NDVIV - NDVIS)2 \qquad (11)$$

The Normalized difference built-up index (NDBI) is for delineating the impervious areas. NDBI was calculated for all the images using SWIR and NIR bands using Eq. (12):

$$NDBI = SWIR - NIR/SWIR + NIR \qquad (12)$$

## Land Surface Temperatures in Different Land Cover Types

After deriving the required parameters like LST, NDVI, and NDBI for all the data sets, their relationships were sought out in various land cover types. In the study area, four different land cover types were selected (residential, industrial, conservation, and desert) for the analysis (Fig. 2). The residential area consists of the residential districts of the city. The land cover in this area is a mixture of impervious surface and vegetation. The ratio of the mixture varies across the residential area.

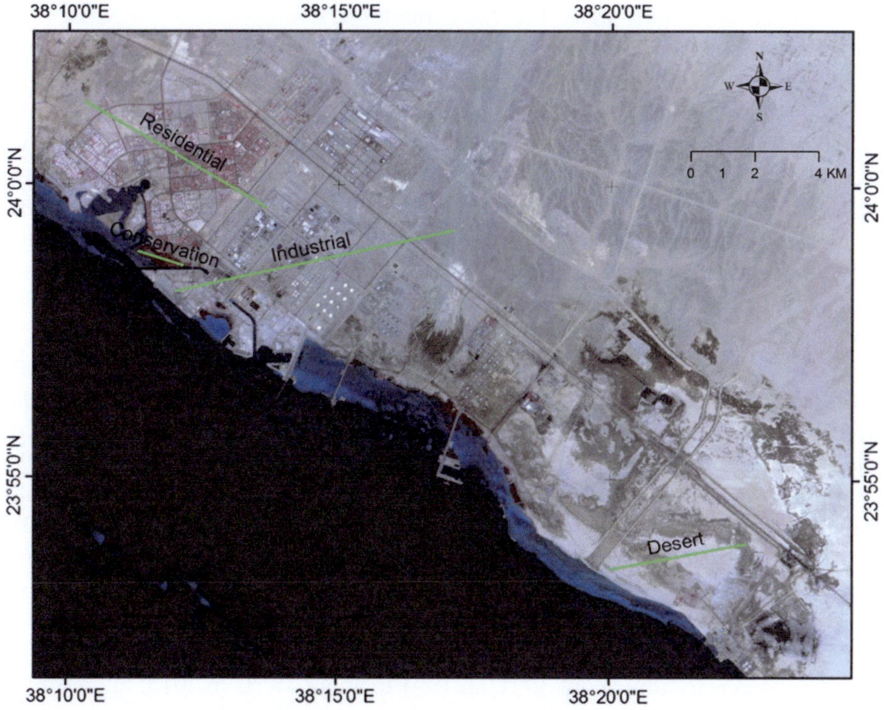


**Fig. 2** Study area with transects across various land covers types

The industrial area includes mainly the industrial areas with little or no vegetation. The conservation area is mainly vegetated areas consisting of mangroves. A desert area at the outskirt of the city was chosen to represent areas with limited anthropogenic activities. Four transects were taken across (from left to right) the four selected land cover types (Fig.2). The land cover types were identified on a high-resolution image (IKONOS) acquired in 2012 and processed as a false color composite. The transects were used to derive the temperature profiles of each land cover type at the different years of study. The NDVI and NDBI ratios calculated for all the images from 1991 to 2015 were compared to the respective LST images to derive their correlations. Random samples of 200 points from each land cover type were used to compute the correlations based on different levels of significance. After that, LST statistics such as minimum, first quartile, median, third quartile, and maximum were computed for each land cover type through box plots. Different graphs and maps were generated from the result by using ArcGISand R software.

## 4 Results and Discussion

The LST images, after the removal of a couple of outliers, show a steady increase in surface temperature from 1991 to 2015, with a marked increase in temperature in the industrial and desert areas. For 1991, the highest temperature is around 37 °C while for 2000, 2007, and 2015 the highest temperatures are 42, 45, and 47 respectively (Fig. 3). Areas with high-density vegetation are relatively cooler including conservation and residential areas. Substantial vegetation cover in the residential part has resulted in the area being cooler than the other parts by around 3–5 °C, and the industrial area is hotter by 5–7 °C (Fig. 4). The thermal regimes of the selected land cover types are depicted in the LST profiles in Fig. 4. The thermal profile of the residential area can be characterized as an urban cool island while that of the industrial part can be described as a UHI (Fig. 4). The profiles of the desert and conservation areas are relatively straight as LST values are constant along the transects.

The results of the correlation between LST, NDVI, and NDBI of all the multi-date images show an inverse relation with NDVI and direct relation with NDBI values in the residential and conservation areas (Table 4). Most of the correlations, especially in the residential area, are significant at the 0.01 level (Table 4). On the other hand, the correlations in the desert and industrial area mainly depict a direct relationship between LST and NDVI because the NDVI values are very low due to very sparse vegetation. Similarly, there is a direct relationship between LST and NDBI. The box plots in Fig. 5 show the temporal changes in LST statistics by land cover types. The plots depict a trend of increasing surface temperatures in the study area. For instance, the median of LST in the industrial area rose from about 30 °C in 1991 to about 38 °C in 2015. Interestingly, while the median value for the desert area dropped in 2015, the values in the other land cover types continue to increase. It might be due to changes in the landscape in

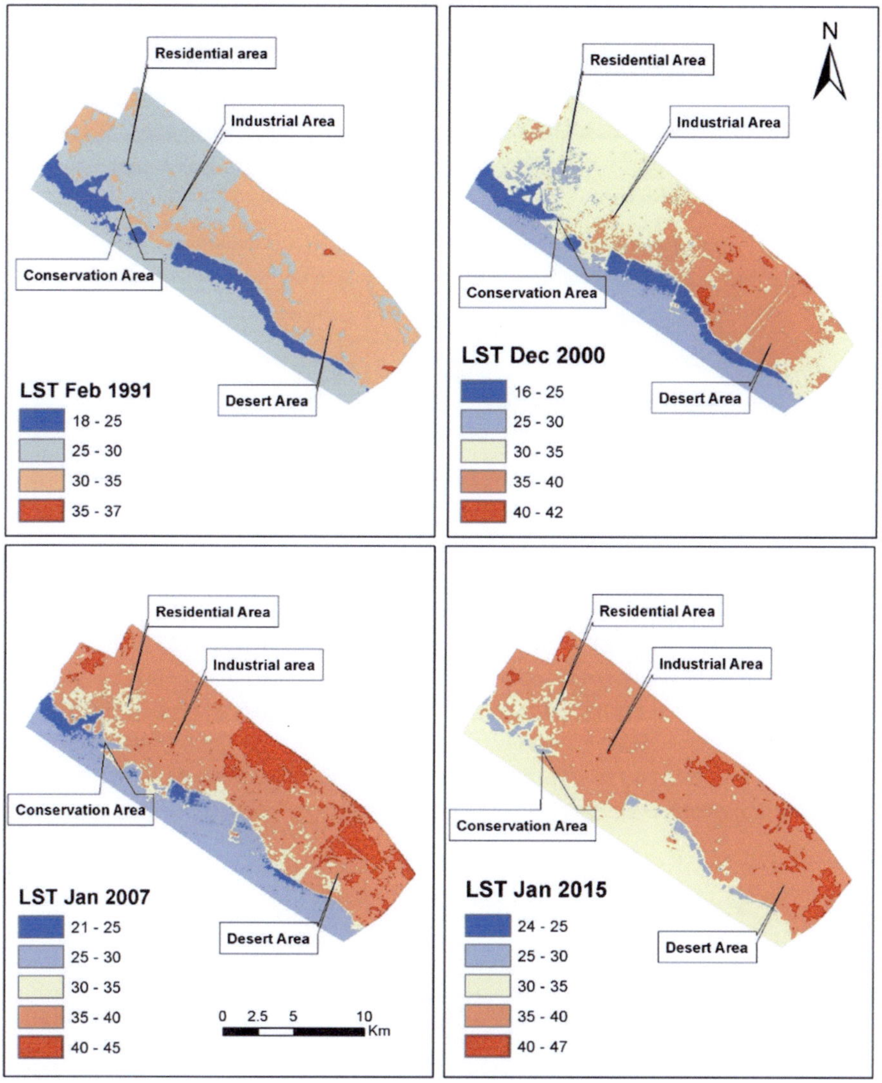


**Fig. 3** Land surface temperature of Yanbu industrial city 1991–2015

the desert area such as an increase in the vegetation cover. Since 2015 is the only year that an inverse relationship between LST and NDVI is depicted for the desert area (Table 4). The box plots also show that the median values in the conservation area increased from about 23 °C in 1991 to about 29 °C in 2015 (Fig. 5). It might probably indicate a depletion of vegetation cover or stress in the vegetation of the conservation area.

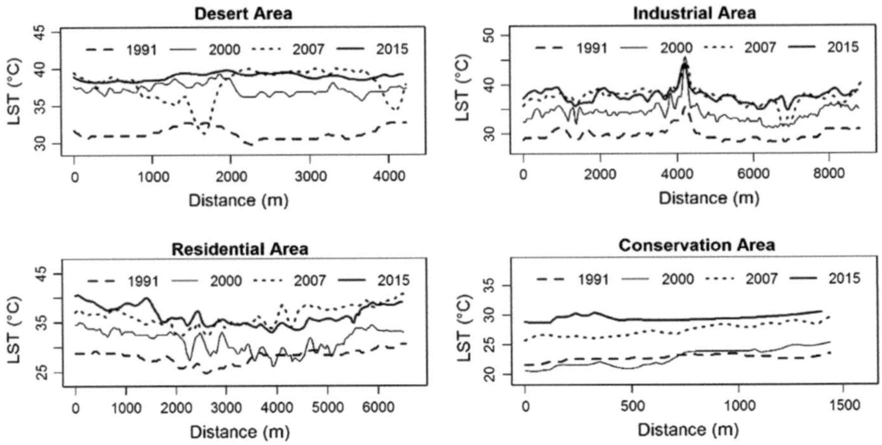


**Fig. 4** LST profile of selected transects by land use/land cover types (1991–2015)

**Table 4** Correlations between LST, NDVI, and NDBI (1991–2015)

| LST/Year | Desert area | | Industrial area | | Residential area | | Conservation area | |
|---|---|---|---|---|---|---|---|---|
| | NDVI | NDBI | NDVI | NDBI | NDVI | NDBI | NDVI | NDBI |
| LST_1991 | 0.46** | 0.64** | −0.60 | 0.08 | −0.33** | 0.66** | −0.13 | 0.60** |
| LST_2000 | 0.25** | −0.08 | 0.26** | 0.30** | −0.56** | 0.67** | −0.37** | 0.47** |
| LST_2007 | 0.28** | 0.57** | 0.11 | 0.08 | −0.36** | 0.57** | −0.04 | 0.57** |
| LST_2015 | −0.39** | 0.18* | 0.29** | 0.19** | −0.36** | 0.58** | −0.75** | 0.78** |

*significant at the 0.05 level; **significant at the 0.01 level

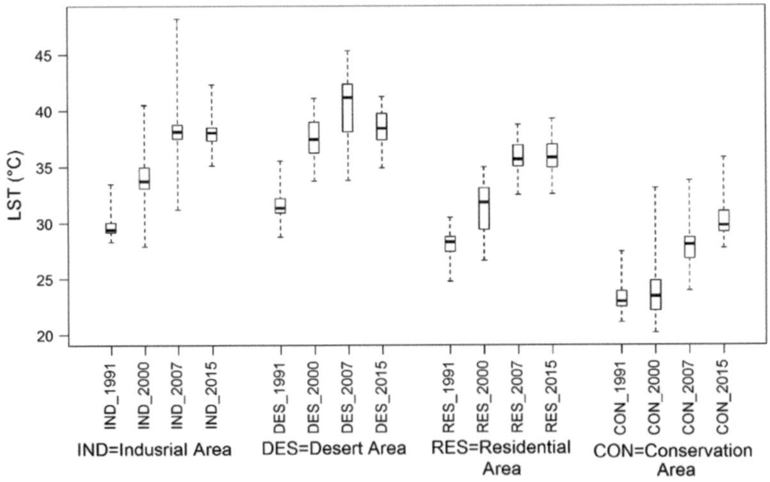


**Fig. 5** Box plot of surface temperature by land use/land cover types (1991–2015)

The heat island effect in the city, especially in the industrial area, can be mitigated by increasing the vegetation cover in the area. Figure 2 reveals that there is virtually no vegetation cover in the industrial area. High industrial activities without mitigating vegetation cover can lead to thermal profiles with temperatures higher than the temperatures of the bare desert as shown in Fig. 4. There is a need to improve the management of the conservation area to avoid degradation and protect the area since it serves as a heat sink and provides ecosystem services to the community. It is also the habitat of aquatic and terrestrial species, and high temperatures due to the degradation of the area might lead to loss of important species especially the coral reefs.

## 5 Conclusions

Satellite remote sensing provide a convenient means and methods to study the dynamics of land surface temperature. The study has explored the capability of Landsat data to identify thermal differences within the urban environment and the possible ways to mitigate the effects of the urban heat island. Analyzing urban thermal regimes is very important as they contribute to global environmental change and directly affect the people living in urban areas. Because the images are of different months and sensors, it cannot be totally ruled out that the observed changes over the years might be due to normal temporal variations or sensors calibrations rather than human-induced factors. Further studies can be carried out by using a greater number of images and time series analysis better to depict temporal variations in the temperature regimes.

**Acknowledgements** The authors wish to thank the USGS EROS Center for making the data used in this study available.

## References

Aina YA, Aleem KF (2014) Assessing the vulnerability of an industrial city to predicted sea level rise using SRTM and GPS observations: the case of Yanbu, Saudi Arabia. Int J Geoinf 10 (3):73–81

Alghamdi AS, Moore TW (2015) Detecting temporal changes in Riyadh's urban heat island. Papers Appl Geogr 1(4):312–325

Barsi JA, Barker JL, Schott JR (2003) An atmospheric correction parameter calculator for a single thermal band earth-sensing instrument. In: Proceedings of IEEE IGARSS, 21–25 July 2003, Toulouse, France, pp 3014–3016

Barsi JA, Schott JR, Palluconi FD, Hook SJ (2005) Validation of a web-based atmospheric correction tool for single thermal band instruments. In: Proceedings of SPIE, Earth observing systems X, vol 5882, Bellingham, WA

Buyantuyev A, Wu J (2010) Urban heat islands and landscape heterogeneity: linking spatiotemporal variations in surface temperatures to land-cover and socioeconomic patterns. Landsc Ecol 25(1):17–33

Carlson TN, Ripley DA (1997) on the relation between NDVI, fractional vegetation cover and leaf area index. Remote Sens Environ 62:241–252

Chander G, Markham BL, Helder DL (2009) Summary of current radiometric calibration coefficients for Landsat MSS, TM, ETM+ and EO-1 ALI sensors. Remote Sens Environ 113:893–903

Chen XL, Zhao HM, Li PX, Yin ZY (2006) Remote sensing image-based analysis of the relationship between urban heat island and land use/cover changes. Remote Sens Environ 104 (2):133–146

Cristobal J, Jimenez-Munoz JC, Sobrino JA, Ninyerola M, Pons X (2009) Improvements in land surface temperature retrieval from the Landsat series thermal band using water vapor and air temperature. J Geophys Res 114:D08103. doi:10.1029/2008JD010616

Deng X, Zhao C, Lin Y, Zhang T, Qu Y, Zhang F, Wang Z, Wu F (2014) Downscaling the impacts of large-scale LUCC on surface temperature along with IPCC RCPs: a global perspective. Energies 7(4):2720–2739

Irish R (2003) Landsat 7 Science Data User Handbook (NASA Goddard Space Flight Centre, Greenbelt, Md., available on-line: http://landsathandbook.gsfc.nasa.gov

Jenerette GD, Harlan SL, Buyantuev A, Stefanov WL, Declet-Barreto J, Ruddell BL, Myint SW, Kaplan S, Li X (2015) Micro-scale urban surface temperatures are related to land-cover features and residential heat related health impacts in Phoenix, AZ USA. Landsc Ecol, 1–16

Jimenez-Munoz JC, Sobrino JA (2003) A generalized single-channel method for retrieving land surface temperature from remote sensing data. J Geophys Res 108(D22):4688. doi:10.1029/2003JD003480

Jimenez-Munoz JC, Cristobal J, Sobrino JA, Soria G, Ninyerola M, Pons X (2009) Revision of the single-channel algorithm for land surface temperature retrieval from landsat thermal-infrared data. IEEE Trans Geosci Remote Sens 47:339–349. doi:10.1109/TGRS.2008.2007125

Li JJ, Wang XR, Wang XJ, Ma WC, Zhang H (2009) Remote sensing evaluation of urban heat island and its spatial pattern of the Shanghai metropolitan area, China. Ecol Complex 6:413–420

Li J, Song C, Cao L, Zhu F, Meng X, Wu J (2011) Impacts of landscape structure on surface urban heat islands: a case study of Shanghai, China. Remote Sens Environ 115:3249–3263

Lim HS, MatJafri MZ, Abdullah K, Saleh NM, AlSultan S (2007) Application of remote sensing for land surface temperature retrieval over Mecca, Defense and Security Symposium, 65410D-65410D, International Society for Optics and Photonics

Lim HS, Jafri M, Abdullah K, Alsultan S (2012) Application of a simple mono window land surface temperature algorithm from Landsat ETM over Al Qassim, Saudi Arabia. Sains Malaysiana 41(7):841–846

RCY [Royal Commission—Yanbu] (2015) Yabu industrial city 4th quarter economic—report #22, RCY, Yanbu

Schowengerdt R (2007) Remote sensing, models and methods for image processing, 3rd edn. Academic Press, London

Schwarz N, Lautenbach S, Seppelt R (2011) Exploring indicators for quantifying surface urban heat islands of European cities with MODIS land surface temperatures. Remote Sens Environ 115(12):3175–3186

Sobrino JA, Raissouni N (2000) Toward remote sensing methods for land cover dynamic monitoring. Application to Morocco. Int J Remote Sens 21:353–366

Sobrino JA, Jimenez-Munoz JC, Paolini L (2004) Land surface temperature retrieval from Landsat TM 5. Remote Sens Environ 90:434–440

Valor E, Caselles V (1996) Mapping land surface emissivity from NDVI: application to European, African and South American areas. Remote Sens Environ 57:167–184

Van De Griend AA, Owe M (1993) On the relationship between thermal emissivity and the normalized difference vegetation index for natural surfaces. Int J Remote Sens 14:1119–1131

Vanos JK, Middel A, McKercher GR, Kuras ER, Ruddell BL (2016) Hot playgrounds and children's health: a multiscale analysis of surface temperatures in Arizona, USA. Landsc Urban Plan 146:29–42

Walawender JP, Szymanowski M, Hajto MJ, Bokwa A (2014) Land surface temperature patterns in the urban agglomeration of Krakow (Poland) derived from Landsat-7/ETM+ data. Pure appl Geophys 171(6):913–940

Yusuf YA, Pradhan B, Idrees MO (2014) Spatio-temporal assessment of urban heat island effects in Kuala Lumpur metropolitan city using Landsat images. J Indian Soc Remote Sens 42 (4):829–837

# The Study of Multi-temporal Analysis of Urban Development and Environmental Changes of the City of Abu Dhabi

**H. Hussein and M. Hussein**

**Abstract** The city of Abu Dhabi is the capital of the United Arab Emirates. It is located on the coast of the Arabian Gulf. Forty years ago the area of Abu Dhabi was composed of nothing but a desert except a small number of buildings. When the Emirates united in the year of 1971, and the late His Highness Sheikh Zayed Bin Sultan Al Nahyan, had his vision of turning the city of Abu Dhabi specifically and the UAE generally into one of the most beautiful countries in the world. Both urban expansion and vegetation bloomed in the country. Abu Dhabi now has one of the highest GDP growth rates in the world, indicating the dramatic increase of urban areas, occupying and building cities, islands, and roads. This project studies the growth of Abu Dhabi city by studying the urban expansion and the percentage of vegetation throughout the years: 1973–2010, using satellite imagery and remote sensing techniques.

**Keywords** Urban expansion · Gross domestic product · Vegetation · Normalized difference vegetation index

## 1 Introduction

The ever-increasing quantity of multi-temporal data provided by the numerous remote sensing satellites that orbit our planet has facilitated the synergistic use of multi-temporal remote sensing data and advanced analysis methodologies resulting in the possibility of solving complex problems related to the monitoring of the Earth's surface and atmosphere. The past 10 years have shown a significant increase

H. Hussein · M. Hussein (✉)
Department of Electrical Engineering,
United Arab Emirates University, Abu Dhabi, UAE
e-mail: mihussein@uaeu.ac.ae



S. Pirasteh and J. Li (eds.), *Global Changes and Natural Disaster Management: Geo-information Technologies*, DOI 10.1007/978-3-319-51844-2_2

in this topic, with plenty of research introducing algorithms to study this flood of data incoming onto our servers. One of the greatest challenges for natural and social scientists in the next decades is to understand how metropolitan areas evolve through the interactions between human behaviors and biophysical processes. The complexity of these interactions is extraordinary. However, our failure to understand and to account adequately for them in policy decisions has historically yielded infrastructure investment and land use decisions with unintended long-term effects (Alberti and Waddell 2000). Previous research has indicated that urban land use/land cover change can cause changes in ecosystem functions and services (Alberti 2005; Grannemann et al. 2000; Issa 2009; Lindenmayer and Fischer 2013; Dai et al. 2011). Many studies have sought to quantify the temporal distribution and arrangement of land cover changes from classified remotely sensed data and to estimate the environmental impact of these changes using landscape metrics. Some study the growth of cities and urban expansion. Others study the atmosphere and environmental changes. This paper uses several satellite photos acquired, to study the growth of vegetation, land, and urban expansion of the capital city of Abu Dhabi in the United Arab Emirates. This city has exhibited a dramatic growth in the short span of 40 years since the 1970s when the seven emirates joined together to form the UAE.

## 2 Study Area and Data Description

The city of Abu Dhabi is in the northeastern part of the Persian Gulf in the Arabian Peninsula. It is on an island less than 250 m (820 ft) from the mainland and is currently joined to the mainland by several bridges. The major land cover/use classes in the area being studied are high-density built-up, low-density built-up, parks/road vegetation. The imagery was acquired from Landsat 5 TM, Landsat 7 ETM, and ALOS-AVNIR depending on their availability over the years of 1973, 1984, 1989, 2001, 2006, and 2010. Two scenes of the ALOS satellite images were necessary for each year in order to obtain full coverage of the area. The following are the images used for studying and comparing.

- 08-19730913-LANDSAT1-MSS-007
- 08-19840429-LANDSAT5-TM-M-070
- 08-19890521-LANDSAT5-TM-M-034
- 08-20010911-LANDSAT7-ETM-M-128
- 08-20060923-ALOS-AVNIR1-195
- 08-20061010-ALOS-ANVIR2-205
- 08-20100704-ALOS-AVNIR-378
- 08-20100721-ALOS-AVNIR-379

# 3 Methodology

## *3.1 Geometric Correction and Mosaicking*

All images used were corrected using standard atmospheric correction usually used for high-resolution satellite imagery. This helps in comparing several multi-temporal satellite images with different times of acquisition taking the atmosphere's effect out (Liu 2013; Liu and Mason 2013). The adjacent ALOS images were mosaicked together using standard stitching. The software ERDAS IMAGINE was used in the scope of this project.

## *3.2 Vegetation Analysis*

Vegetation has a very unique spectral analysis with two major peaks (for reflection) one in the NIR band and another in the green band. On the other hand, the spectral analysis of vegetation shows high absorption in the red band. Therefore to investigate the vegetation areas in an image, an NDVI—Normalized Difference Vegetation Index, is used.

$$\mathrm{NDVI} = \frac{\mathrm{NIR} - \mathrm{Red}}{\mathrm{NIR} + \mathrm{Red}} \tag{1}$$

The NDVI is a difference formula that then uses the division to normalize the outputted values giving an output histogram range from 0 to 1. Equation 1 was adjusted a bit to make the vegetation the only thing visible in the image while all other pixels were set to 'null'.

$$\mathrm{If}\left(\frac{\mathrm{NIR} - \mathrm{Red}}{\mathrm{NIR} + \mathrm{Red}}\right) > 0.2\,\mathrm{then}\,255,\ \mathrm{else}\,\mathrm{Null}$$

Contrast enhancement is not required in this case since the outputted image shows the vegetation with color and anything else in the image is not shown at all (null).

## *3.3 Urban Expansion Analysis*

In urban expansion manually selected areas were taken to identify the high and low-density built-up areas of the city. Once chosen, default software statistics were run.

### 3.4 *Classification*

The maximum likelihood algorithm was used to classify the images, training pixels were chosen for the classification of each land cover, and LANDSAT7 in 2001 was selected for land cover classification, this image represented the majority of variance in the data.

## 4 Results and Discussion

### 4.1 *Vegetation Analysis*

Inputting Eq. 1 mentioned earlier and applying this algorithm to the images provided, showed the dramatic increase of vegetation in the city of Abu Dhabi through the 30 or so years being analyzed. Figure 1 graphically shows the foliage available in the city in 1973. The scarcity of vegetation in the area can be highly noted, the merging of the vegetation highlight with the original LANDSAT image shows us where this vegetation is growing, and this can be seen in Fig. 2.

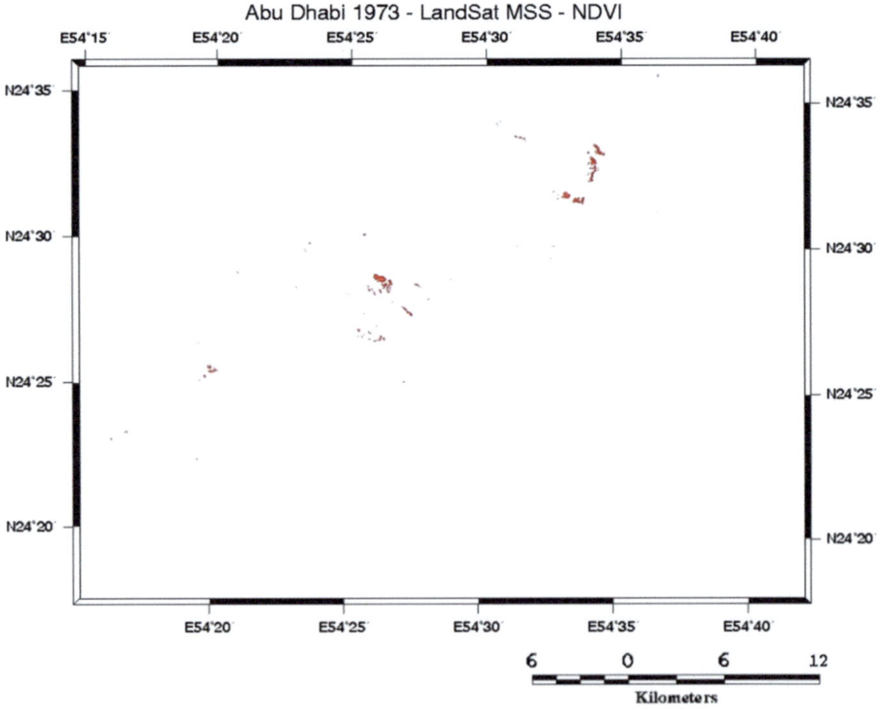


**Fig. 1** Vegetation in Abu Dhabi—1973

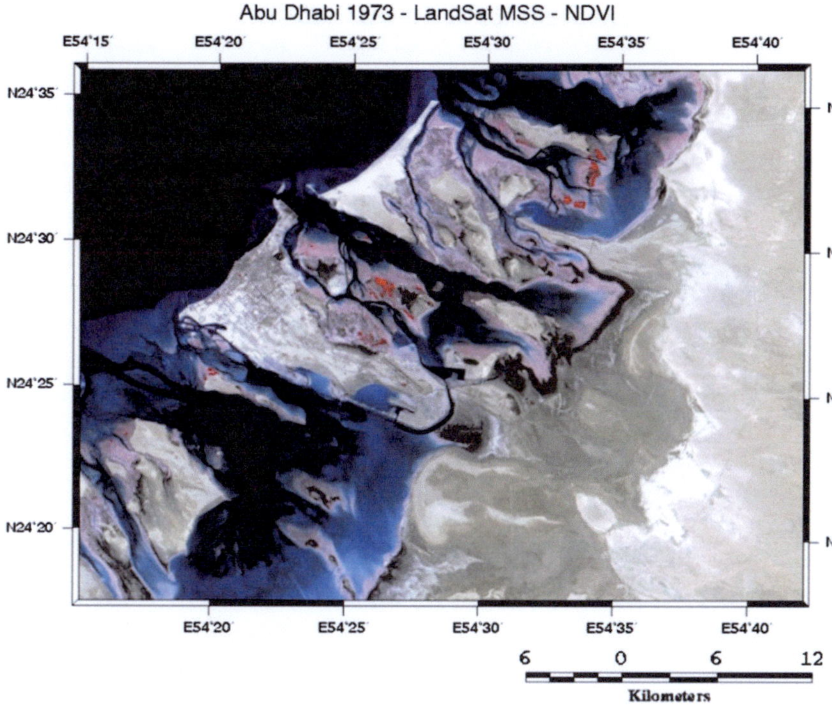


**Fig. 2** Vegetation in Abu Dhabi—merged image

Looking closely it can be seen that the highlighted areas are actually in shallow water bodies, from here it can be noted that the foliage in Abu Dhabi consisted of mangroves. Mangroves are forests of salt-tolerant trees that grow in tidal waters of around the coast of Abu Dhabi. These plants are the only naturally growing vegetation in the city area.

Similar analysis was applied to all images mentioned above. For comparison, Fig. 3 shows the vegetation in 2010 of the same area.

Looking at the statistics of the two images shown above we can see that the vegetation areas in the city of Abu Dhabi have increased drastically, from 438.477 acres in 1973 to 18,279.81 acres in 2010. Figure 4 shows a graph of all the statistical data of the vegetation in Abu Dhabi between the years 1973 and 2010. It has to be clearly noted here that all vegetation increase in this city is because of human planting and watering, the UAE by nature is a desert area and plants do not flourish this dramatically without help.

A slight change in the trend in the year 2001 shown in the plot in Fig. 4, in which the vegetation did not show as much increase as the previous years, noted by the change in slope in the graph. This is due to urban expansion where in some cases to expand urban areas some of the vegetation had to be removed. Once all urban areas had been planned and mapped out, vegetation was added. Looking at

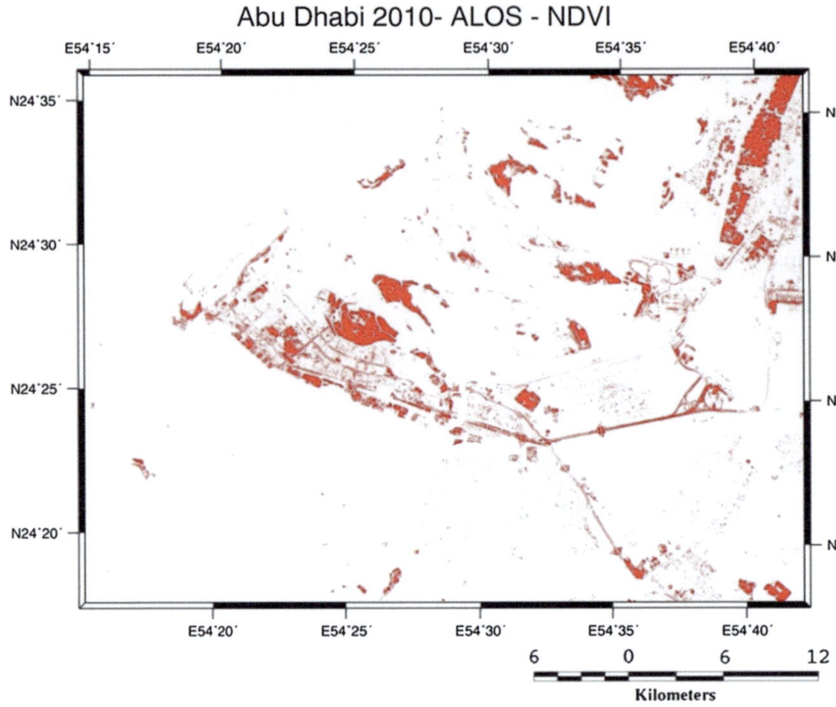


**Fig. 3** Vegetation in Abu Dhabi—2010

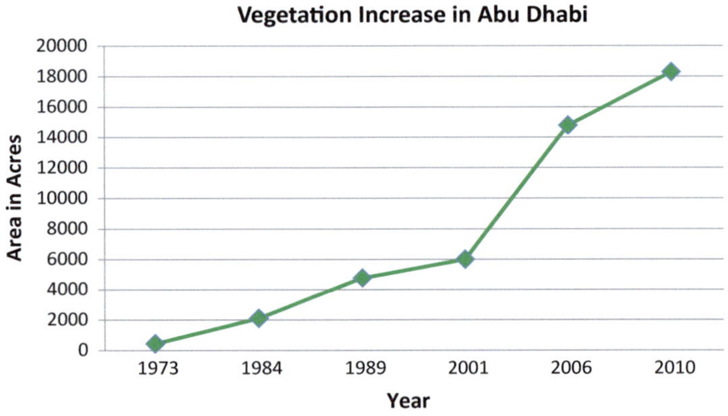


**Fig. 4** Statistical data of vegetation area in Abu Dhabi

the graph further, we can see a sudden boom in the vegetation in the year 2006, increasing by almost 10,000 acres in 5 years. This then slows back down in 2010. To explain this phenomenon further, the urban areas mapping has to be done and the results compared with those of vegetation increase.

It should be noted here that almost all of the foliage seen in the city is planted and watered by man, meaning the city municipality has placed a lot of resources, manpower, money, water, and time, to better beautify the city. Usually in vegetation trends around the world, studies notice a decrease in vegetation with urban expansion. Abu Dhabi has been wise to understand the importance of planting these trees, and growing green parks for a better, healthier ergonomic city.

## 4.2 Urban Expansion

To investigate urban expansion in the area of Abu Dhabi city, the areas were manually chosen, labeled and the statistics for them calculated. Figure 5 shows the urban area in Abu Dhabi, in the year 1973. The selected region shows a low populated area. Looking at the image for the year 1973 there are no highly populated areas. This can be related to the statistical data of the population of Abu Dhabi shown in Table 1. In the year 1973 only 125,933 (Grannemann et al. 2000) people occupied a rough estimate of 10,025.56 acres of land, hence the low populated distribution.

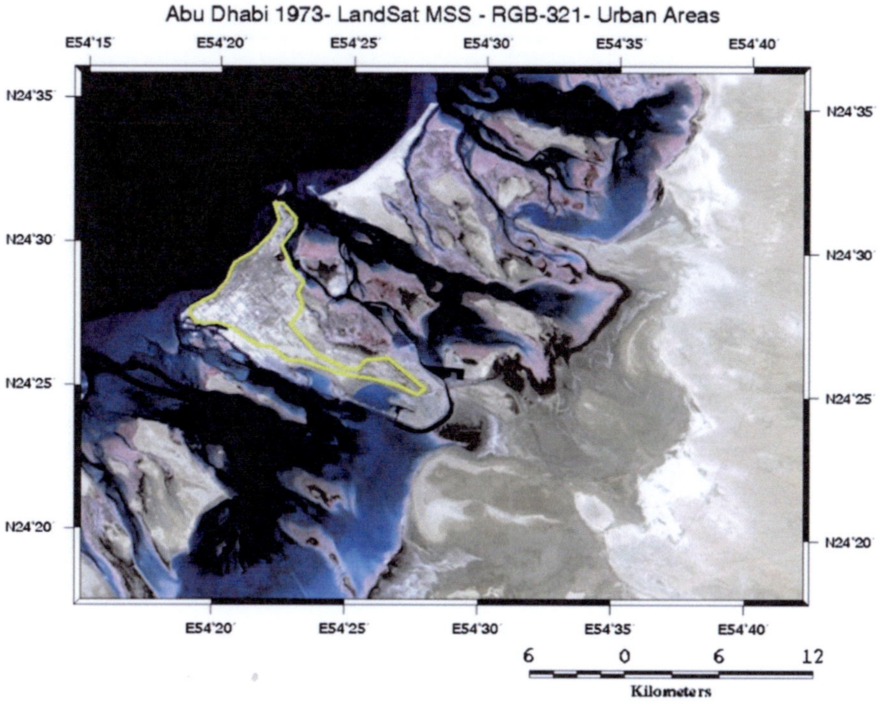


**Fig. 5** Urban area in Abu Dhabi—1973

**Table 1** Population of Abu Dhabi (Grannemann et al. 2000)

| Year | Population |
|---|---|
| 1973 | 125,933 |
| 1978 | 309,422 |
| 1980 | 420,455 |
| 1984 | 523,181 |
| 1989 | 678,348 |
| 1994 | 874,633 |
| 1999 | 1,071,141 |
| 2001 | 1,155,963 |
| 2006 | 1,461,479 |
| 2010 | 1,967,659 |

In comparison, the year 2010 had a total population of 1,967,659 (Grannemann et al. 2000) densely occupying 35,979 acres of land and lightly occupying 59,346 acres of land. The lightly occupied areas are mainly mapped out and being built for future expansion purposes. The urban area mapping of the city in 2010 can be seen in Fig. 6, where the yellow regions represent highly populated areas, and the green regions represent low populated areas.

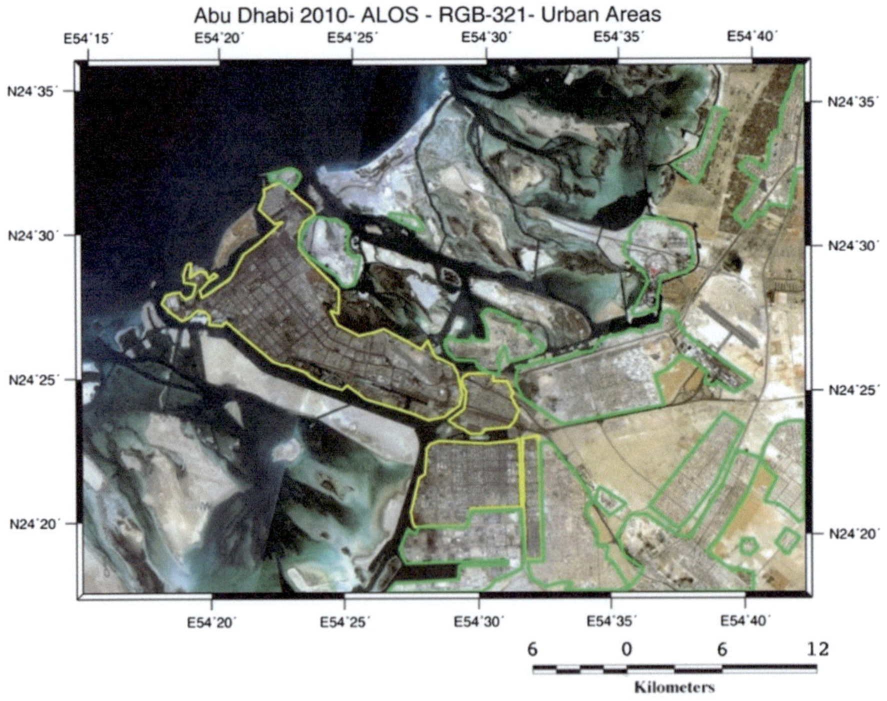


**Fig. 6** Urban area in Abu Dhabi—2010

To further analyze urban expansion, a graph of the statistical data of both the highly and low populated urban areas was made shown in Fig. 7. Looking at this graph and tracking the change of low populated areas to high populated areas, the following changes have occurred between the years 1973 and 1984: the entire low populated area occupied in 1973 was expanded and built further to occupy more residents changing the label of the region to a highly populated area (all images can be found in the appendix section). A couple of other areas have been very lightly occupied on the mainland, off of the island of Abu Dhabi. The trend of highly populated areas can be seen in Fig. 7 as follows, there is a dramatic increase between the years of 1973 and 1984, and there is then a slightly less dramatic increase between the years 1984 and 2001. To relate the population increase Fig. 8 was plotted and the slopes compared. Considering that the slope of the population

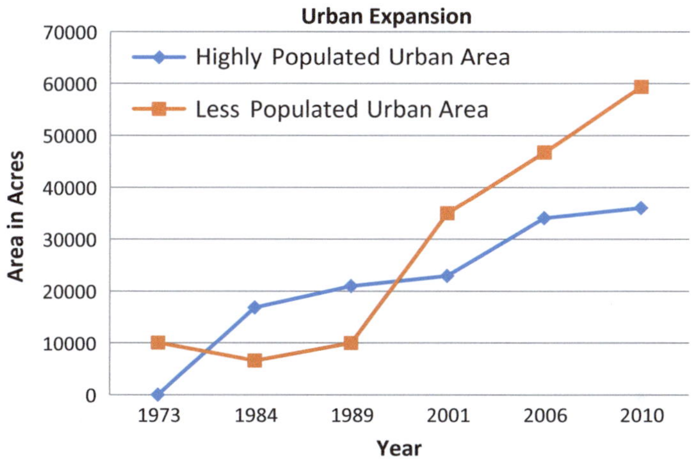


**Fig. 7** Urban expansion graph

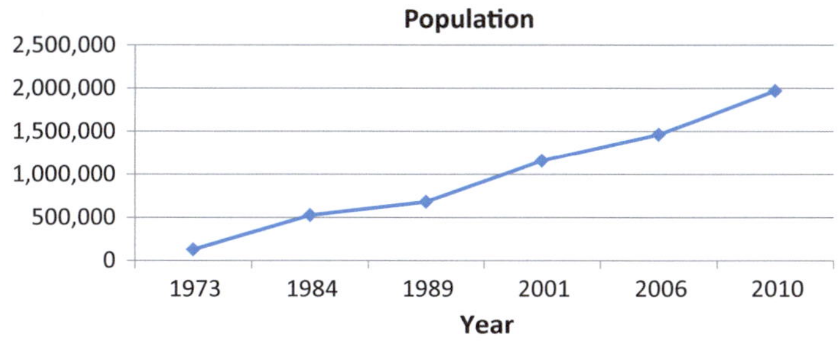


**Fig. 8** Graph of the population

increase between the years 1984 and 1989 is almost the same as that between the years 1973 and 1984, the further population increase seemed to have moved into areas farther out and off the island which can be seen as the increase in the number of lowly populated areas. The trend can be followed in the same way for the rest of the graph explaining how the city of Abu Dhabi expanded over the time of 40 years.

## 4.3 *Classification*

Due to the fact that there were many different types of classes in the region, the following classes were taken when choosing training areas: highly populated urban areas, low populated urban area, vegetation, shallow water, deep sea water, and eight barren land areas highlighting the difference in texture and content of the land's sand. Figure 9 shows this classification, the legend shows the different classes chosen and identified.

It can be noted that the class that is mostly seen in the city in 2001 is the low populated area class.

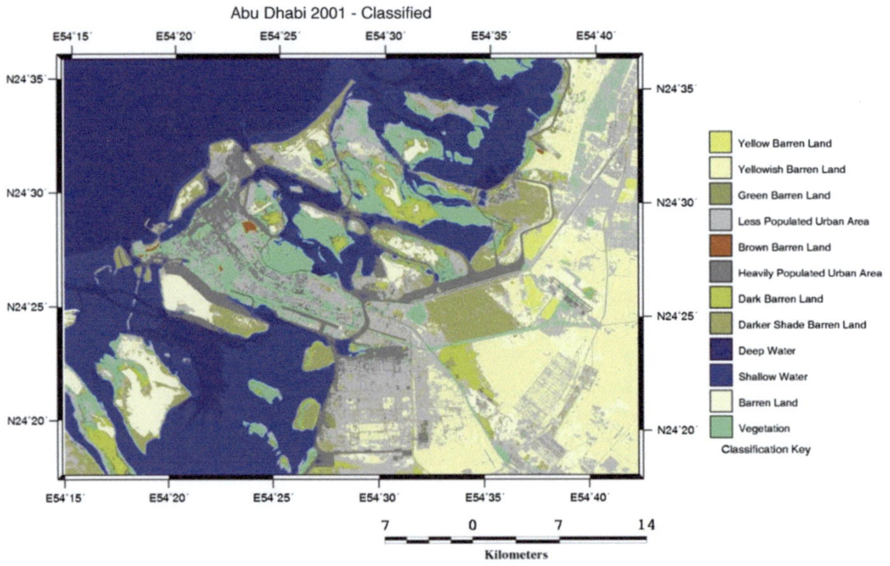


**Fig. 9** Classified image of Abu Dhabi

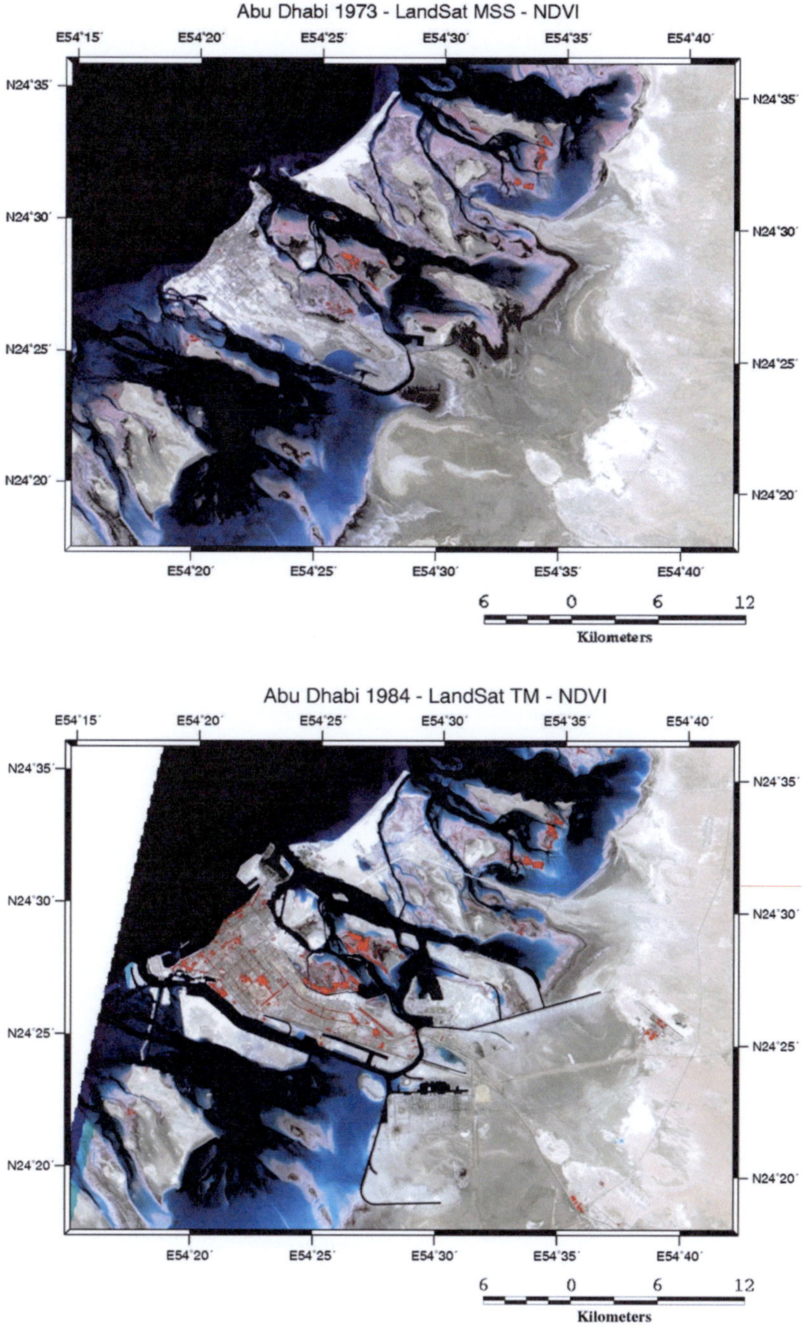


Fig. 10 NDVI and RGB-432 overlapped for vegetation

## *4.4 Relating the Plots*

In order to further study the vegetation/urban expansion, their relative plots are to be compared in details. The comparison will be done in segments between each two corresponding years.

**The year 1973–1984**

The population increase was first looked at and an increase of roughly 397,000 can be estimated within this time frame. As mentioned earlier, the entire lightly populated area in 1973 became a highly populated area in 1984. While little areas of the light population were built to accommodate and future expansion. The vegetation has increased in this period, considering the UAE had become an officially united country in 1972 and started expanding and building a city. Figure 10 shows the vegetation in years 1973 and 1984 respectively. The figures are shown in an RGB-432 fake color composite to highlight vegetation in red. It was then overlapped with the NDVI red and white images mentioned earlier, to make the vegetation a solid red. It can be seen that the city put more effort into planting and making grass parks in the year 1984.

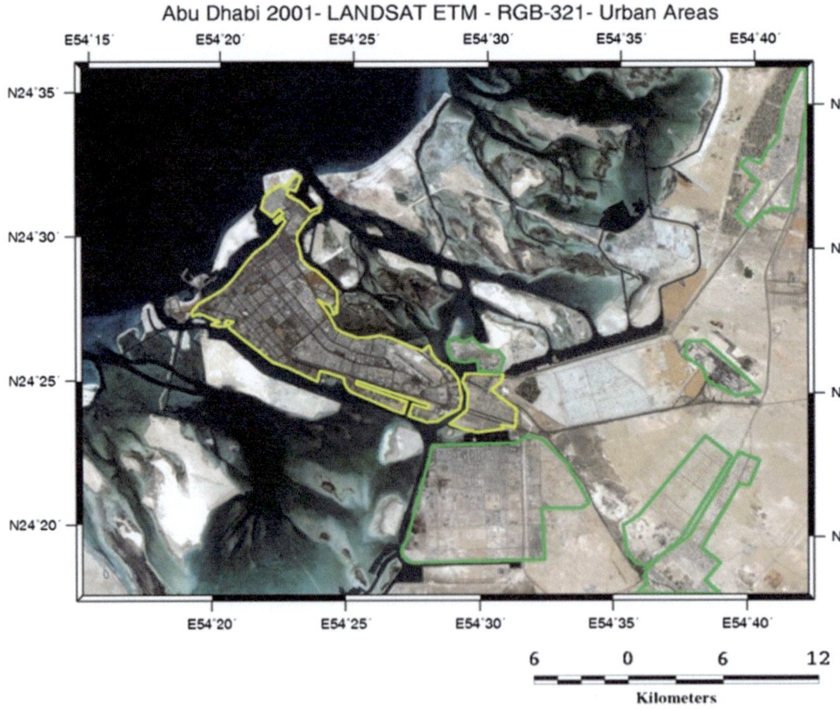


**Fig. 11** Urban areas in Abu Dhabi 2001

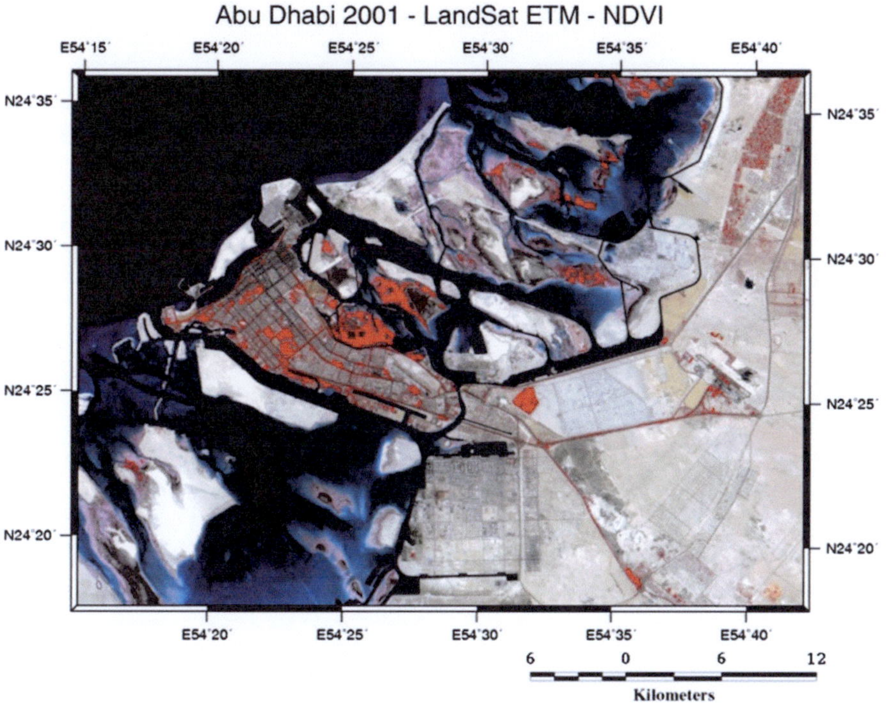


**Fig. 12** NDVI and RGB-432 overlapped for vegetation

## The year 1989–2001

The population increase in this time frame can be roughly estimated to 477 thousand. This is a very high increase when compared to the previous year, and would demand a huge expand in the urban area. However, looking at the area of the highly populated region there is not much of an increase—except for a small piece of land on the mainland into the connected islands (North-West)—but on the other hand, the low populated area shows a drastic increase. Looking at the relevant Figs. 11 and 12, we can see in details how each region has expanded. Statistically, the vegetation has increased with a meager 1000 acres, that can be seen in some farmland areas in the northeast, the mangroves shown with a very dense concentration in the center, and the Abu Dhabi golf club is shown as a patch of heavy vegetation on the mainland. However if we look more closely, we can see that some of the vegetation on the main roads of the city has been removed. This can explain the very small change in the overall area of the vegetation, where it increased on one side but decreased in the other.

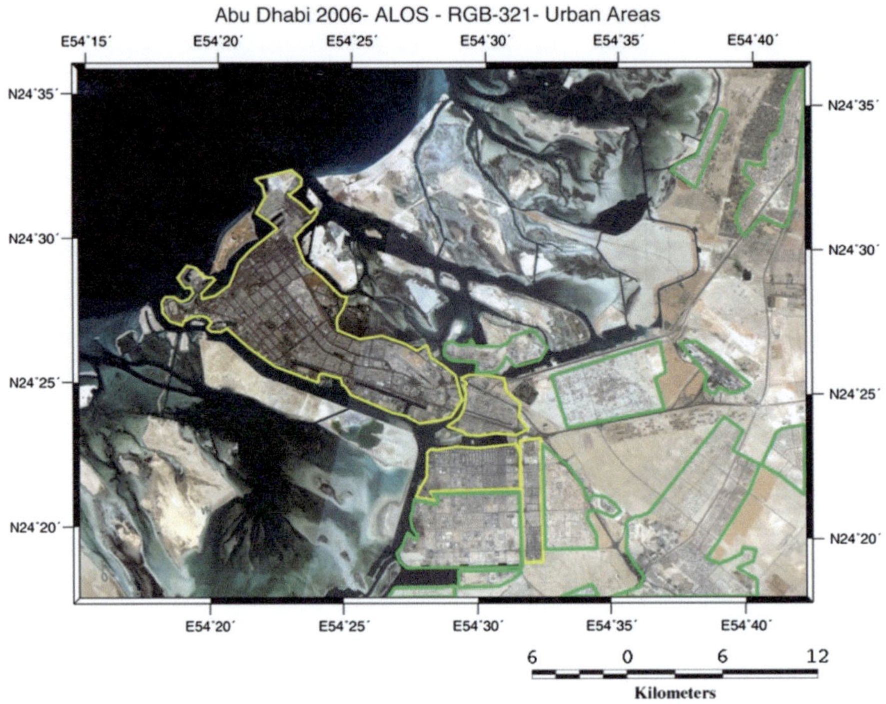


**Fig. 13** Urban areas in Abu Dhabi 2006

## The year 2001–2006

The population remains at a steady increase of 305 thousand, which is close to the amount of increase in the previous time frame, 1989–2001 when the urban area expansion is rather large, and we can see the regions selected in Fig. 13 are covering almost all the land in 2006. Vegetation has increased at about 8000 acres in this time frame. Looking at Fig. 14 that shows the vegetation in 2006 we can see the huge increase between this year and 2001 in Fig. 12 both in the main city and the farmlands to the northeast. A few of the lightly populated regions show beginning signs of vegetation as can be seen in Fig. 14.

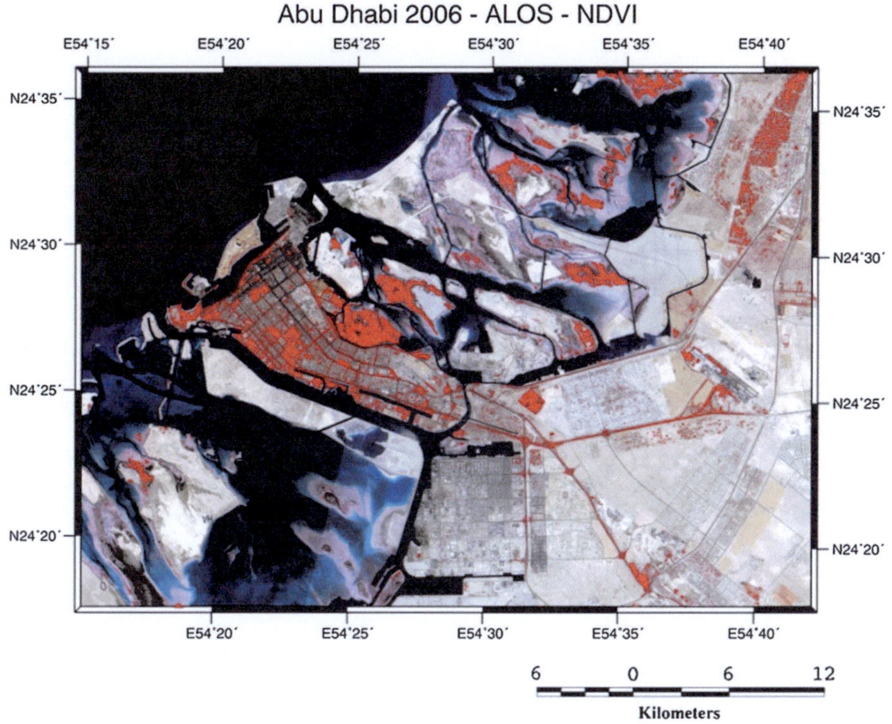


**Fig. 14** NDVI and RGB-432 overlapped for vegetation

## 5 Conclusion

This research demonstrates the use of multi-temporal satellite images to map spatial and temporal patterns of urban and vegetation expansion.

Vegetation changes were observed and the huge difference in green areas between the year 1984 and 2001 was noted. Also, the trend of the vegetation was to increase the vegetation areas inside the Abu Dhabi Island and on the highway streets which leads to Dubai city.

In 2001, Abu Dhabi built partially new islands such as Lulu, Sadiat, Yas, and Al Reem Islands, due to this urban expansion was observed and the huge difference in the infrastructure between the year 1984 and 2001 was also commented on. Also, Abu Dhabi has shifted the concentration of the buildings to new cities like Madinat Sheikh Khalifa A and B on the Dubai road and Madinat Sheikh Mohammad Bin Zayed on the Al Ain road. Finally, this project successfully investigated the main changes in the urban areas of Abu Dhabi and the consequential environmental changes.

## References

Alberti M (2005) The effects of urban patterns on ecosystem function. Int Reg Sci Rev 28(2):168–192

Alberti M, Waddell P (2000) An integrated urban development and ecological simulation model. Integr Assess 1(3):215–227

Dai Y, Zhang X, Yan C, Huijuan A, Zhu C (2011) Monitoring coastal zones land use and land cover changes using remote sensing: a case study of Jiangsu, China. In: 2011 19th international conference on geoinformatics, IEEE, pp 1–4

Grannemann NG, Hunt RJ, Nicholas JR, Reilly TE, Winter TC (2000) The importance of ground water in the great lakes. USGS Water Resour Invest. http://water.usgs.gov/ogw/pubs/WRI004008/WRIR_00-4008.pdf. Accessed 20 Mar 2012

Issa SM (2009) Land development assessment on the preserved Al Sammalyah island/UAE using multi-temporal aerial photographs and GIS. Sensors 5:8–13

Lindenmayer DB, Fischer J (2013) Habitat fragmentation and landscape change: an ecological and conservation synthesis. Island Press

Liu JG (2013) Remote sensing for earth and science applications and GIS, course portfolio

Liu JG, Mason PJ (2013) Essential image processing and GIS for remote sensing. Wiley, Hoboken

N.p. (2016) Web. 28 May 2016. http://www.geoinformatic.org/projects/remote-sensing-30-years-of-development-and-vegetation-change-in-abu-dhabi/

SCAD – Home. Scad.ae. http://www.scad.ae/en/Pages/default.aspx

# CRF-Based Simultaneous Segmentation and Classification of High-Resolution Satellite Images

**Weihong Cui, Guofeng Wang, Chenyi Feng, Yiwei Zheng and Jonathan Li**

**Abstract** Scale selection and uncertainty of image segmentation is still an intractable problem which influences the image classification results directly. To solve this problem, we adopt a CRF (Conditional Random Field)-based method to do segmentation and classification simultaneously. In this method, using probabilistic graphical model, we construct a three-level potential function which includes the pixels, the objects, and the link among the pixels and the objects to model their relations. We transform it to an optimization problem and use the graph cut algorithm to get the optimal solution. This method can refine the segmentation while getting good classification result. We do some experiments on the GF-1 high spatial resolution satellite images. The experiment results show that it is an effective way to improve the classification accuracy, avoid the boring segmentation scale and parameters selection and will highly improve the efficiency of image interpretation.

**Keywords** Object-oriented · High resolution image · Segmentation · Classification · CRF

W. Cui (✉)
School of Remote Sensing and Information Engineering,
Wuhan University, Wuhan, China
e-mail: whcui@whu.edu.cn; w28cui@uwaterloo.ca

W. Cui
Collaborative Innovation Center for Geospatial Technology, Wuhan, China

W. Cui · J. Li
Mobile Mapping Lab, University of Waterloo, Waterloo, Canada
e-mail: junli@uwaterloo.ca

G. Wang
China Highway Engineering Consulting Corporation, Beijing, China

C. Feng · Y. Zheng
Xi'an University of Science and Technology, Xi'an, China
e-mail: 491016316@qq.com

Y. Zheng
e-mail: 276014260@qq.com

 
S. Pirasteh and J. Li (eds.), *Global Changes and Natural Disaster Management: Geo-information Technologies*, DOI 10.1007/978-3-319-51844-2_3

## 1 Introduction

LULC (Land Use and Land Cover) information extraction is a key step of national geographic state surveying and monitoring. To get more detailed information of the LULC, about ten first level classes, 46 second level classes, and some third level classes are proposed by NASG of China. Abundant features are offered by high spatial resolution images, at the same time, large intra-class variations and low interclass variation exist in them. The situation makes even more challenge on high spatial resolution images processes and uses. The familiar eCongnition software has offered a series of object-oriented image classification tool, while because of the uncertainty of segmentation which makes it is very difficult to be used in the surveying project. So, in real applications, the operators should use high-resolution remote sensing images such as GF-1 or UAV images and do heavy manual interpretation work. Many facts proved that the segmentation problem is still very intractable. Researchers have proposed lots of image segmentation methods, such as MST (Felzenszwalb and Huttenlocher 2004), mean shift (Comaniciu and Meer 2002), watershed (Beucher and Meyer 1992), graph cut (Boykov and Jolly 2001), etc. Each of these segmentation methods has to set the segmentation parameters which may lead to over-segmentation or under-segmentation. How to improve the segmentation accuracy? Can we combine the segmentation with classification together to get the optimized segmentation boundaries while improving the classification accuracy? Machine learning and computer vision technology make it possible. Accumulation of manual interpretation results can provide massive training samples which are very useful resource for learning the segmentation and classification model. Multi-scale features and context features are very important clues to recognize the objects. In this paper, we adopt Conditional Random Field (CRF), which can model these features and relationship of among objects to realize the simultaneous segmentation and classification.

## 2 Related Work

In computer vision and object recognition field, we could divide its processes into two types based on its outputs. One is called object detection, which gives the center position and a rectangle box of the target. For example, face recognition (Déniz et al. 2011), human detection (Zhu et al. 2006), part-based model (Felzenszwalb et al. 2009), sparselet model (Song et al. 2012), bag-of-features model (Lazebnik et al. 2006; Yang et al. 2009), sparse coding (Yang et al. 2009; Gao et al. 2010; Jia et al. 2012; Jiang et al. 2012), deep learning (Krizhevsky et al. 2012; Donahue et al. 2013), and so on. These methods can get the center positions and bounding boxes of the targets in the image. The other one is called semantic image segmentation, which is a process of simultaneous segmentation and recognition of an input image into regions and their associated categorical labels. An effective way to achieve this goal is to assign a label to each pixel of the input image and set some structural constraints on the output label space. Graph is a great tool to model the relations among different objects. With the MRF and MAP (Maximum Posterior Probability) developments, MRF and CRF which use probability graph model to represent this type of problems has been used widely. They model the label problem based on pixel features (Lafferty et al. 2001; Blake et al. 2004; Shotton et al. 2006; Larlus and Jurie 2008; Toyoda and Hasegawa 2008; Gould et al. 2008), region features (Yang et al. 2007; Kohli and Torr 2009; Fulkerson et al. 2009; Tighe and Lazebnik 2010; Yang and Forstner 2011b), multi-level regions (Russell et al. 2009; Schnitzspan et al. 2009; Yang et al. 2010; Kohli et al. 2013; Ladicky et al. 2014). In which, the image classification problem was expressed as a potential function and was transformed to an energy optimal problem, and they can get the boundary of the targets. Yang and Forstner (2011a, b), Zhong and Wang (2007), and Montoya-Zegarra et al. (2015) use CRF to model spatial and hierarchical structures to label and classify images of man-made scenes, such as buildings, roads, etc., and demonstrate the effective of CRF method in extracting man-made targets from high-resolution remote sensing images. We also could find that the common point of these two types of target recognition methods are both represent and make use of many kinds of features, such as SIFT (Scale Invariant Feature Transform), HOG (Histogram of Oriented Gradients), LBP (Local Binary Patterns), etc.

In remote sensing image applications, we usually need to get the accurate boundary of the targets. With this request, we take the second type method and the features used in the first type into consideration to get the object class label and boundary. In this paper, to make use of the spectral, texture and context features, we construct a three-level potential function which includes pixel level, segment level, and up-down-layer level with reference to Yoyoda and Hasegawa (2008), Kohli and Torr (2009), Ladicky et al. (2014), and use graph cut (Boykov et al. 2001; Boykov and Jolly 2001) to find the optimal solution. The experiments on GF-1 satellite image with 2 m spatial resolution showed that the proposed method is an efficient way to improve the segmentation and classification accuracy.

## 3 CRF-Based Image Classification Method

Features in different scales are very helpful information for image interpretation. High-resolution remote sensing images provided more detailed textures. To utilize the different scales and different kinds of features, we select four kinds of popular features and three scales' segmentation results to describe the classes. Figure 1 shows the workflow. We calculate the SIFT, LBP, Texton, Color SIFT features of each image. To reduce the computing complexity and to realize sparse representation, we apply the k-means clustering on four kinds of features that were got in the trained images, respectively, to get the visual words. Then construct the pixel and different scales segments potentials. Here, we select the mean shift image

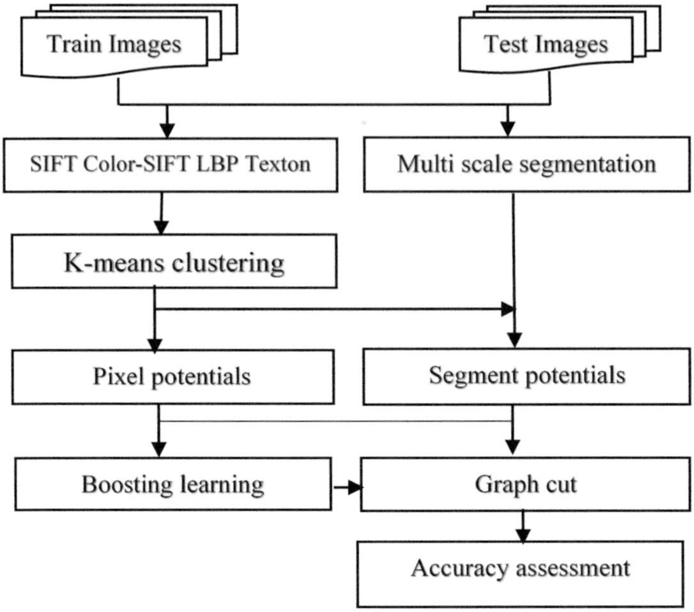


**Fig. 1** Workflow of our method

segmentation method. Through boosting learning we got the classifiers. Finally, combine the pixel level, segment level, and the different level potentials together and use graph cut optimize algorithm to find the finest solution which means the fittest class of each pixel.

## *3.1 Features Calculation*

### 3.1.1 SIFT Descriptor

In computer vision, SIFT descriptor are calculated after getting the key points and are used to realize point match generally. As the SIFT descriptor has the capability of describing the spatial distribution of a window, it has been used in many target detection research (Lazebnik et al. 2006; Yang et al. 2009; Jia et al. 2012), which just labeled the rectangle region where the object may exist.

This feature is derived from a $4 \times 4$ gradient window by using a histogram of $4 \times 4$ samples per window in 8 directions. The gradients are then Gaussian weighted around the center. This leads to a 128-dimensional feature vector. It reflects the distribution of gradients' direction. Before using this feature, we should normalize its value.

### 3.1.2 Color-SIFT Descriptor

Color is an important component for districting objects. Color invariant descriptors are proposed to increase illumination invariance and discriminative power. There are many different methods to obtain color descriptors, Van De Sande et al. (2010) compared the invariance properties and the distinctiveness of color descriptors. In this paper, we choose RGB-SIFT descriptor to describe the color invariant. For the RGB-SIFT descriptor, SIFT descriptors are computed for every RGB channel independently.

### 3.1.3 LBP Feature

An LBP is a local descriptor that captures the appearance of an image in a small neighborhood around a pixel. Due to its discriminative power and computational simplicity, LBP texture operator has become a popular approach in various applications. An LBP is a string of bits, with one bit for each of the pixels in the neighborhood. Each bit is got by thresholding the neighborhood of each pixel with the value of the center pixel. Here we select a $3 \times 3$ neighborhood, it is a string of 8 bits and has 256 possible LBPs.

### 3.1.4 Texton Feature

The term texton was proposed by Julesz (1981) first. It is described as "the putative units of pre-attentive human texture perception" (Julesz 1981). Leung and Malik (2001) use this term to describe vector quantized responses of a linear filter bank. Textons have been proven effective in categorizing materials as well as generic object classes. In this paper, we select three Gaussians, four Laplacian of Gaussians (LoG) and four first-order derivatives of Gaussians to build the filter bank. The three Gaussian kernels (with $\sigma = 0.1, 0.2, 0.4$) are applied to each CIE $L$, $a$, $b$ channel, the four LoGs (with $\sigma = 0.1, 0.2, 0.4, 0.8$) were applied to the $L$ channel only, and the four derivatives of Gaussians were divided into the two $x$- and $y$-aligned sets, each with two different values of $\sigma$ ($\sigma = 0.2, 0.4$). Derivatives of Gaussians were also applied to the $L$ channel only. Thus produced 17 final filter responses. Therefore, each pixel in each image has associated a 17-dimensional feature vector (Julesz 1981).

## *3.2 Mean Shift Image Segmentation*

Mean shift is a robust feature space analysis approach and can delineate arbitrarily shaped clusters in it. Mean shift-based image segmentation has been widely used in many kinds of image, including high-resolution remote sensing images. The segmentation is actually a merging process performed on a region that is produced by the mean shift filtering. It considers both spatial domain and spectral domain while merging. For both domains, the Euclidean metric is used. Because the Euclidean distance in RGB color space does not correlate well to perceived difference in color by people, we use the LUV color space which better models the perceived difference in color in this space Euclidean distance. The use of the mean shift segmentation algorithm requires the selection of the bandwidth parameter $h$ = (hr, hs), which determines the resolution of the mode detection by controlling the size of the kernel. To get a different scale segment result, we select 3 group of parameters which are $h1$ = (3.5, 3.5), $h2$ = (5.5, 3.5), $h3$ = (3.5, 5.5).

### 3.3 Potential Function

Image classification is a problem of assigning an object category label to each pixel in a given image. MRFs are the most popular models to incorporate local contextual constraints in labeling problems. Let $Ii$ be the label of the $i$th site of the image set $S$, and $Ni$ be the neighboring sites of site $i$. The label set $L\,(=\{l_i\}_{i\in S})$ is said to be a MRF on $S$ w.r.t. a neighborhood $N$ iff the following condition is satisfied

$$P\big(l_i|l_{S-\{i\}}\big) = P(l_i|l_{Ni}) \tag{1}$$

Let $l$ be a realization of $L$, then $P(l)$ has an explicit formulation (Gibbs distribution):

$$P(l) = \frac{1}{Z}\exp(-\frac{1}{T}E(1)) \tag{2}$$

$$E(l) = \sum_{c\in C} V_C(l) = \sum_{\{i\}\in C_1} V_1(l_i) + \sum_{\{i,i'\}\in C_2} V_2(l_i, l_{i'}) + \cdots \tag{3}$$

where $E(l)$ is the energy function, $Z$ is a normalizing factor, called the partition function, $T$ is a constant, Clique $C_k = \{\{i, i', i'', \cdots\}|i, i', i'', \ldots$ are neighbors to one another$\}$. $V_C(l)$ is the potential function, which represent a priori knowledge of interactions between labels of neighboring sites. Maximizing a posterior probability is equivalent to minimizing the posterior energy:

$$L^* = \arg\min_L E(L|X) \tag{4}$$

Let $G$ = ($S$, $E$) be a graph, then ($X$, $L$) is said to be a CRF if, when conditioned on $X$, the random variables $l_i$ obey the Markov property with respect to the graph:

$$P\left(l_i|X, l_{S-\{i\}}\right) = P(l_i|X, l_{N_i}) \tag{5}$$

where $S\{i\}$ is the set of all sites in the graph except the site $i$, $N_i$ is the set of neighbors of the site $i$ in $G$. We can find that CRF can directly infer posterior $P(L|X)$. In CRF, the potentials are functions of all the observation data as well as that of the labels. The CRF allows us to incorporate shape, color, texture, layout, and edge cues in a single unified model using a conditional potential. CRF model can be used to learn the conditional distribution over the class labeling given an image. Some kinds of the CRF have been proposed, for example, the image pixels (Gould et al. 2008; Toyoda and Hasegawa 2008), patches (Yang et al. 2007; Fulkerson et al. 2009; Kohli and Torr 2009; Tighe and Lazebnik 2010), or a hierarchy of regions (Russell et al. 2009; Yang et al. 2010; Kohli et al. 2013). We use a CRF model (Kohli and Torr 2009; Russell et al. 2009; Ladicky et al. 2014) to learn the conditional distribution over the class labeling given an image. We define the conditional probability of the class labels $L$ given an image $X$

$$E(X) = \sum_{i \in V} \theta_v \varphi_i(x_i) + \sum_{(i,j) \in \varepsilon} \theta_\varepsilon \varphi_i\left(x_i, x_j\right) + \sum_{c \in S} \theta_s \varphi_c(X_c) \tag{6}$$

where $V$ is a set of the image pixels, $\varepsilon$ is the set of edges in an 8-connected grid structure; $S$ is a set of image segments, $\varphi_i(x_i)$, $\varphi_{ij}\left(x_i, x_j\right)$ and $\varphi_c(X_c)$ are the potentials defined on them, $\theta_v$, $\theta_\varepsilon$ and $\theta_s$ are the model parameters, and $i$ and $j$ index pixels in the image, which correspond to nodes in the graph. In this paper, we defined three potentials which are unary potential, pairwise potential, and region potential. We will describe these potentials as follows.

### 3.3.1 Unary Potential

The unary potential allows for local and global evidence aggregation, each potential models the evidence from considering a specific image feature. Usually, it is computed from the color of the pixel and the appearance model for each object. However, color alone is not a very discriminative feature and fails to produce accurate segmentations and classification. This problem can be overcome by using sophisticated potential functions based on color, texture, location, and shape priors. The unary potential used by us can be written as

$$\varphi_i(x_i) = \theta_s \varphi_s(x_i) + \theta_l \varphi_l(x_i) + \theta_t \varphi_t(x_i) + \theta_{cs} \varphi_{cs}(x_i) \tag{7}$$

where $\theta_s$, $\theta_l$, $\theta_t$ and $\theta_{cs}$ are parameters weighting the potentials obtained from SIFT, LBP, texton, and color SIFT respectively.

### 3.3.2 Pairwise Potential

The pairwise potentials have the form of a contrast sensitive Potts model (Kohli and Torr 2009).

$$\varphi_{ij}(x_i, x_j) = \begin{cases} 0 & \text{if } x_i = x_j \\ g(i,j) & \text{otherwise} \end{cases} \tag{8}$$

where the function $g(i,j)$ is an edge feature based on the difference in colors of neighboring pixels (Song et al. 2012). It is typically defined as

$$g(i,j) = \theta_p + \theta_v \exp(-\theta_\beta ||I_i - I_j||^2) \tag{9}$$

where *Ii* and *Ij* are the color vectors of pixel *i* and *j* respectively. $\theta_p$, $\theta_v$ and $\theta_\beta$ are model parameters whose values are learned using training data.

### 3.3.3 Region Consistency Potential

The region consistency potential is modeled by the robust Pn Potts model (Kohli and Torr 2009). It supports all pixels belonging to a segment taking the same label and allows some variables in the segment to take different labels and reflect the consistency of segments. It is very useful in obtaining object segmentations with fine boundaries. We refer the reader to Kohli and Torr (2009) for more details. It takes the form of

$$\varphi_c(X_c) = \begin{cases} N_i(X_c)\frac{1}{Q}\gamma_{\max} & \text{if } N_i(X_c) \le Q \\ |c|^{\theta_\alpha}(\theta_p^h + \theta_v^h G(c)) & \text{otherwise} \end{cases} \tag{10}$$

where $N_i(X_c) = \min_k(|c| - n_k(X_c))$, which denotes the number of variables in the clique c not taking the dominant label. $\gamma_{\max} = |c|^{\theta_\alpha}(\theta_p^h + \theta_v^h G(c))$, and Q is the truncation parameter which controls the rigidity of the higher order clique potential. *G(c)* is used to evaluate the consistency of all constituent pixels of a segment, the variance of the response of a unitary was used, that is

$$G(c) = \exp\left(-\theta_\beta^h \frac{||\sum_{i\in c}(f(i) - \mu)^2||}{|c|}\right) \tag{11}$$

where $\mu = \frac{\sum_{i\in c} f(i)}{|c|}$ and $f()$ is a function being used to evaluate the quality of a segment. This enhanced potential function gives rise to a cost that is a linear truncated function of the number of inconsistent variables (Kohli and Torr 2009).

In this paper, we use boosting algorithm to train the three part of the energy function. The boosting algorithm helps us select features and get a strong classifier.

### *3.4 Graph Cut*

Given the CRF model and its learned parameters, we wish to find the most probable labeling *l*, i.e., the labeling that minimize the energy function of (6). The graph cut-based $\alpha$-expansion and $\alpha\beta$-swap is an effective way to solve energy minimization problem (Boykov et al. 2001). It transforms the energy minimizing problem to min-cut of graph problem. It has been successfully used to minimize energy functions composed of pairwise potential functions.

As a kind of move making algorithms, the first step of graph cut is initializing nodes, edges and build up the graph. In this paper, in accordance with the constituents of the energy function, the nodes and edges consist of the pixel, pairwise, and three scales segments-level components. The three levels' potentials are calculated and used to initialize the edge weight. The initial label image is set according to the minimum cost of each class. Then, it computes optimal alpha-expansion moves for labels in some order, accepting the moves only if they increase the objective function. The algorithm's output is a strong local maximum, which means the solution of minimum energy was found.

## 4 Results and Discussion

In our experiments, we select high-resolution satellite images to verify our method's availability. The satellite image was the merged image of the multispectral image and panchromatic image of 2 m spatial resolution of "GF-1." The test area is located in the southeast of Liaoning province in China. The image was manually interpreted, which include paddy field, dry land, bare land, shrub land, buildings, and greenhouse. Some area is labeled to "other class." The image has $20000 \times 8000$ pixels. We split this large image into lots of 128 * 128 small images which are then divided into the train and test group and take the 30% of the small images as train images. To compare the result with the other object-oriented image classification method, we select the wide used software eCognition to compare. In eCognition classification process, through many trials, we selected the scale parameter 100, calculated the mean value and standard variation of each band, the secondary angle moment of GLCM feature, and use the Nearest Neighbor algorithm to realize the object-oriented image classification. Figure 2 shows the origin image (a), ground truth (b), classification result of this paper's method (c), and classification result of eCognition. Table 1 gives the image classification accuracies of these two methods. From them, we could find the accuracy of our method is higher than the traditional object-oriented image classification. At the same time, compared with the traditional method, the phenomena of salt-and-pepper was greatly improved.

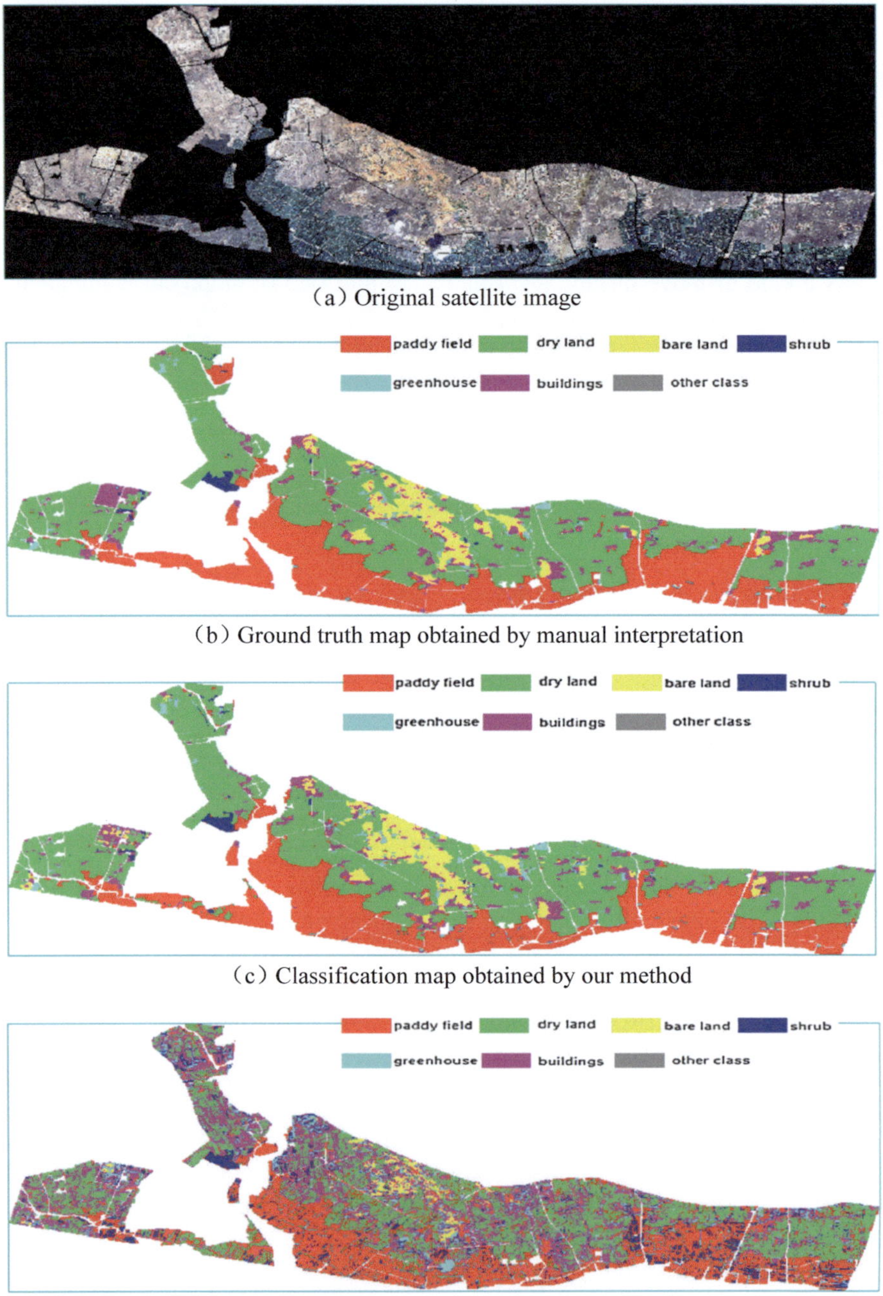


(a) Original satellite image

(b) Ground truth map obtained by manual interpretation

(c) Classification map obtained by our method

(d) Classification map obtained by eCognition

**Fig. 2** Original satellite image, ground truth, and classification results

**Table 1** Comparison of classification accuracy

| | Paddy field | Dry land | Bare land | Shrub | Building | Green house | Others | Overall accuracy | Average accuracy | Kappa |
|---|---|---|---|---|---|---|---|---|---|---|
| eCognitioan | 0.7943 | 0.8256 | 0.7534 | 0.7456 | 0.8304 | 0.6021 | 0.8912 | 0.8076 | 0.7775 | 0.6712 |
| Our method | 0.9303 | 0.9420 | 0.8014 | 0.8392 | 0.9045 | 0.7328 | 0.9366 | 0.9108 | 0.8695 | 0.8564 |

## 5 Conclusion

From the experiment procedures and results, we could find that this method has three main valuable aspects to be used in image classification. First, it does not need manual segmentation which could avoid the boring segmentation scale selection problem. Second, it can make good use of the existing classification results to train the classifier. Third, it can realize self-selecting features and improve the segmentation and classification results. It may be a way of large of samples-based image interpretation. At the same time, we must admit that this method is still a time-consuming process. How to use it to the task of large-scale geographical conditions general survey and monitoring is still a problem to be solved.

**Acknowledgements** The study was partially supported by the High-resolution Comprehensive Traffic Remote Sensing Application program under Grant No. 07-Y30B10-9001-14/16, the National Natural Science Foundation of China under Grant No. 41101410 and Foundation of Key Laboratory for National Geographic State Monitoring of National Administration of Survey, Mapping and Geoinformation under Grant No. 2014NGCM.

## References

Beucher S, Meyer F (1992) The morphological approach to segmentation: the watershed transformation. Mathematical morphology in image processing. Marcel Dekker, New York, pp 433–481

Blake A, Rother C, Brown M, Perez P, Torr P (2004) Interactive image segmentation using an adaptive GMMRF model. Computer vision-ECCV 2004. Springer, Berlin, pp 428–441

Boykov YY, Jolly MP (2001) Interactive graph cuts for optimal boundary and region segmentation of objects in ND images. In: Proceedings of eighth IEEE international conference on computer vision, 2001, ICCV 2001, vol 1. IEEE, New York, pp 105–112

Boykov Y, Veksler O, Zabih R (2001) Fast approximate energy minimization via graph cuts. IEEE Trans Pattern Anal Mach Intell 23(11):1222–1239

Comaniciu D, Meer P (2002) Mean shift: a robust approach toward feature space analysis. IEEE Trans Pattern Anal Mach Intell 24(5):603–619

Déniz O, Bueno G, Salido J, De la Torre F (2011) Face recognition using histograms of oriented gradients. Pattern Recogn Lett 32(12):1598–1603

Donahue J, Jia Y, Vinyals O, Hoffman J, Zhang N, Tzeng E, Darrell T (2013) Decaf: a deep convolutional activation feature for generic visual recognition. arXiv preprint arXiv:1310.1531

Felzenszwalb PF, Huttenlocher DP (2004) Efficient graph-based image segmentation. Int J Comput Vision 59(2):167–181

Felzenszwalb PF, Girshick RB, McAllester D, Ramanan D (2009) Object detection with discriminatively trained part-based models. IEEE Trans Pattern Anal Mach Intell 32(9):1627–1645

Fulkerson B, Vedaldi A, Soatto S (2009) Class segmentation and object localization with superpixel neighborhoods. In: International conference on computer vision, vol 9, pp 670–677

Gao S, Tsang IWH, Chia LT (2010) Kernel sparse representation for image classification and face recognition. Computer Vision–ECCV 2010. Springer, Berlin, pp 1–14

Gould S, Rodgers J, Cohen D, Elidan G, Koller D (2008) Multi-class segmentation with relative location prior. Int J Comput Vision 80(3):300–316

Jia Y, Huang C, Darrell T (2012) Beyond spatial pyramids: Receptive field learning for pooled image features. In: 2012 IEEE conference on computer vision and pattern recognition (CVPR), IEEE, New York, pp 3370–3377

Jiang Z, Zhang G, Davis LS (2012) Submodular dictionary learning for sparse coding. In: 2012 IEEE conference on computer vision and pattern recognition (CVPR), IEEE, New York, pp 3418–3425

Julesz B (1981) Textons, the elements of texture perception, and their interactions. Nature 290 (5802):91–97

Kohli P, Torr PH (2009) Robust higher order potentials for enforcing label consistency. Int J Comput Vision 82(3):302–324

Kohli P, Osokin A, Jegelka S (2013) A principled deep random field model for image segmentation. In: Proceedings of the IEEE conference on computer vision and pattern recognition, pp 1971–1978

Krizhevsky A, Sutskever I, Hinton GE (2012) Imagine classification with deep convolutional neural networks. In: Advances in neural information processing systems, pp 1097–1105

Ladicky L, Russell C, Kohli P, Torr PH (2014) Associative hierarchical random fields. IEEE Trans Pattern Anal Mach Intell 36(6):1056–1077

Lafferty J, McCallum A, Pereira FC (2001) Conditional random fields: probabilistic models for segmenting and labeling sequence data. In: Proceedings of the eighteenth international conference on machine learning (ICML ‘01), San Francisco, CA, USA, pp 282–289

Larlus D, Jurie F (2008) Combining appearance models and markov random fields for category level object segmentation. In: IEEE conference on computer vision and pattern recognition, 2008, CVPR 2008. IEEE, New York, pp 1–7

Lazebnik S, Schmid C, Ponce J (2006) Beyond bags of features: Spatial pyramid matching for recognizing natural scene categories. In: 2006 IEEE computer society conference on computer vision and pattern recognition, vol 2. IEEE, New York, pp 2169–2178

Leung T, Malik J (2001) Representing and recognizing the visual appearance of materials using three-dimensional textons. Int J Comput Vision 43(1):29–44

Montoya-Zegarra JA, Wegner JD, Ladický L, Schindler K (2015) Semantic segmentation of aerial images in urban areas with class-specific higher-order cliques. ISPRS Ann Photogramm Remote Sens Spat Inf Sci 2(3):127–133

Russell C, Kohli P, Torr PH (2009) Associative hierarchical crafts for object class image segmentation. In: 2009 IEEE 12th international conference on computer vision. IEEE, New York, pp 739–746

Schnitzspan P, Fritz M, Roth S, Schiele B (2009) Discriminative structure learning of hierarchical representations for object detection. In: IEEE conference on computer vision and pattern recognition, pp 2238–2245

Shotton J, Winn J, Rother C, Criminisi A (2006) Textonboost: joint appearance, shape and context modeling for multi-class object recognition and segmentation. Computer vision–ECCV 2006. Springer, Berlin, pp 1–15

Song H, Zickler S, Althoff T, Girshick R, Fritz M, Geyer C, Felzenszwalb P, Darrell T (2012) Sparselet models for efficient multiclass object detection. In: European conference on computer vision, pp 802–815

Tighe J, Lazebnik S (2010) Superparsing: scalable nonparametric image parsing with superpixels. Computer vision–ECCV 2010. Springer, Berlin, pp 352–365

Toyoda T, Hasegawa O (2008) Random field model for integration of local information and global information. IEEE Trans Pattern Anal Mach Intell 30(8):1483–1489

Van De Sande KE, Gevers T, Snoek CG (2010) Evaluating color descriptors for object and scene recognition. IEEE Trans Pattern Anal Mach Intell 32(9):1582–1596

Yang M, Forstner W (2011b) Regionwise classification of building facade images. In: Photogrammetric image analysis, LNCS 6952, Springer, Berlin, pp 209–220

Yang MY, Förstner W (2011a) A hierarchical conditional random field model for labeling and classifying images of man-made scenes. In: 2011 IEEE international conference on computer vision workshops (ICCV workshops), IEEE, New York, pp 196–203

Yang L, Meer P, Foran DJ (2007) Multiple class segmentation using a unified framework over mean-shift patches. In: IEEE conference on computer vision and pattern recognition, 2007, CVPR'07. IEEE, New York, pp 1–8

Yang J, Yu K, Gong Y, Huang T (2009) Linear spatial pyramid matching using sparse coding for image classification. In: IEEE conference on computer vision and pattern recognition (CVPR), 2009. IEEE, New York, pp 1794–1801

Yang M, Forstner W, Drauschke M (2010) Hierarchical conditional random field for multi-class image classification. In: International conference on computer vision theory and applications, pp 464–469

Zhong P, Wang R (2007) A multiple conditional random fields ensemble model for urban area detection in remote sensing optical images. IEEE Trans Geosci Remote Sens 45(12): 3978–3988

Zhu Q, Yeh MC, Cheng KT, Avidan S (2006) Fast human detection using a cascade of histograms of oriented gradients. In: 2006 IEEE computer society conference on computer vision and pattern recognition, vol 2. IEEE, New York, pp 1491–1498

# The Dynamic of Dike-Pond System in the Pearl River Delta During 1964–2012

**Yuenan Li, Kai Liu, Yang Liu and Yuanhui Zhu**

**Abstract** As a traditional mode of agricultural production in the Pearl River Delta (PRD) region, the dike-pond systems (DPS) play an essential role in local economic development and the ecology–agriculture balance. Since 1964, evolution of the systems has been influenced by climate change, policy, and the market economy. We used multi-temporal remotely sensed data (KH–4 data from 1964, KH–9 data from 1976, Landsat TM data from 1988, Landsat ETM+ data for 2000, and HJ–1A data for 2012) to extract land use information for the PRD region. We used this information to describe the development of dike-pond systems in the PRD region and discuss the driving mechanisms in the evolution of the systems.

**Keywords** Pearl River Delta region · Dike-pond system · Remote sensing · LULC change

## 1 Introduction

The dike-pond systems (DPS) are artificial wetlands, which are influenced mainly by reciprocity between ponds and the surrounding land. The DPS are developed by the local people of south China to make full use of available land and water resources (Gongfu 1982; Zhong and Wang 1993; Korn 1996). Since the 1980s, the evolution of dike-pond systems in the Pearl River Delta (PRD) region of South China has occurred without scientific planning, and this has consequently resulted

Y. Li (✉)
Department of Geography and Environmental Management,
University of Waterloo, Waterloo, Canada
e-mail: liyuenan.sysu@gmail.com

K. Liu · Y. Zhu
School of Geography and Planning, Sun Yat-sen University, Guangzhou, China

Y. Liu
Department of Geography, University of Cincinnati, Cincinnati, USA

 
S. Pirasteh and J. Li (eds.), *Global Changes and Natural Disaster Management: Geo-information Technologies*, DOI 10.1007/978-3-319-51844-2_4

in serious eco-environmental problems. These difficulties attracted the attention of researchers, who began to focus on repair and renovation of the dike-pond systems, as well as eco-environmental quality assessments, and other issues (Nie and Luo 2003; Li and Luo 2005).

With the development of remote sensing techniques, more researchers began to describe the dike-pond systems in the PRD region. Wang and Kang (1997) researched the dike-pond systems in Shunde and Nanhai using TM4 data for 1984 and TM5 data for 1994. He and Guan (1998) monitored and compared the dike-pond systems in Shunde, Nanhai District, and Zhongshan City via multi-temporal and multisensor satellite image data, including TM4 data for 1985 and SPOT data for 1995. Yee (1999) described the distribution of the dike-pond system in Shunde using multi-temporal satellite images (TM4 data for 1984 and TM5 data for 1994). Liu and Wang (2008) monitored the spatial-temporal dynamic change of dike-pond systems in Foshan City using TM5 data from 1988, 1998, and 2006.

However, recent studies of dike-pond systems in the PRD region have chosen to combine statistical analysis with field investigation in small regions. Thus, the research is deficient in some aspects. For example, many of the studies lack timeliness and representativeness. Most of the analysis has used qualitative approaches since it is difficult to quantitatively describe the distribution dike-pond systems. Although some researchers started to address these problems using remote sensing techniques in the later half of 1990s, several gaps in the research were still evident.

The most serious problem was that the research periods were short due to the unavailability of early research data and the obsolete collection methods of current research efforts. The earliest remotely sensed data used in this field were mainly the Landsat TM4 imagery of the early 1980s, which was restricted by existing remote sensing techniques and other objective factors. It was rare to find research that represented the spatial distribution patterns of dike-pond systems in the PRD region during that period. More recently, Wang and Xia (2011) used Landsat TM remotely sensed data to analyze the dike-pond systems of Nanhai City.

Remote sensing data provide a means to qualitatively and quantitatively discuss and analyze the distribution patterns, dynamic change processes, characteristics, and driving forces of dike-pond systems. This type of analysis can represent a fundamental scientific basis for the healthy development of the systems in every city in the PRD region, and provide a model for China in terms of further development of sustainable approaches to modern agriculture.

We selected nine cities in the PRD region for our study area. We described the spatial distribution of the dike-pond systems for the period 1964–2012 by integrating remote sensing techniques and utilizing multisource remotely sensed data. We further analyzed and discussed the evolution and dynamic changes in the dike-pond systems in the PRD region, and then discussed the factors that influence land use change in the dike-pond region using both qualitative and quantitative methods.

## 2 Study Area

The PRD region is also called the PRD economic circle, and it comprises nine cities, including: Guangzhou, Shenzhen, Dongguan, Foshan, Jiangmen, Zhuhai, Zhongshan, Zhaoqing, and Huizhou. Figure 1 shows the location. The Guangdong Provincial Party Committee and Provincial Government first suggested creation of the PRD economic region in 1994, and its successful development was mainly due to investment by wealthy Hong Kong businessmen, who also took important leadership roles in the project. This region represents a pioneer zone for Chinese

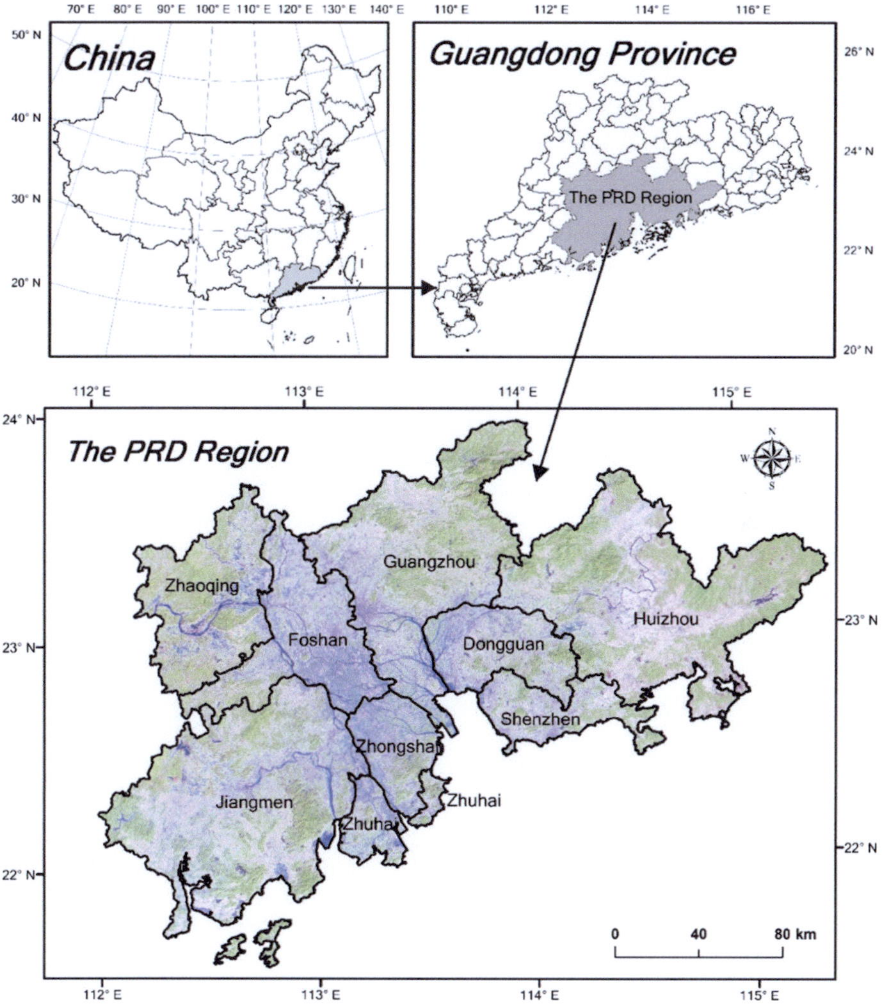


**Fig. 1** Location of the Pearl River Delta region

reform and the "opening-up" policy, and is also a central region in Chinese ecological development. Thus, it possesses a strategic position in China's economic and social development.

The study area has a subtropical monsoon climate. Average annual precipitation in the region ranges from 1600 to 2300 mm, with a flood season from April to September, when 81–85% of the precipitation of the entire year falls. Typhoons are regular here, and mainly occur from July to September.

Dike-pond systems have existed in the PRD region since the Tang Dynasty, and after the late Song Dynasty, the systems were most important part of the economy of the region for several hundred years (Zhong and Wang 1993; Yee 1999). By the middle of the Qing dynasty, the dike-pond industry had developed into significant economic force. Dike-pond systems reached their peak in 1920s in terms of distribution, and the systems greatly contributed to economic prosperity and social development (Lo 1996). Since 1949, the government of the People's Republic of China has conducted comprehensive planning in relation to the dike-pond industry and has intensified water conservancy projects to spread the use of advanced agricultural technology. In the mid-twentieth century, the dike-pond systems were mainly located in the middle of the PRD region, including Shunde, Nanhai, Xinhui, and other sites. Since the 1980s, the level of industrialization in the PRD region has increased greatly and urbanization has also sped up, which has led to environmental pollution and degradation of agricultural production. Hence, the area of the traditional dike-pond system has sharply decreased, and a great deal of dike-pond land has been reclaimed for urbanization (Nie and Li 2001; Guo and Xu 2011). As a result, the dike-pond lands have become increasingly fragmented and isolated. The dike-pond systems in the deltas, estuaries, coastal areas, and other relatively developed regions were most affected. Meanwhile, due to the lack of government management of these systems in recent years, dike-ponds have decreased in productivity, accumulated contaminants, and experienced eutrophication and other serious problems. Consequently, they have begun to threaten human health (Nie and Luo 2003; Guo and Xu 2011).

# 3 Methodology

## *3.1 Overview of Workflow*

This study aimed to comprehensively analyze the development and evolution of dike-pond systems in the PRD region from 1964 to 2012. The technical methodology used in this study is shown in Fig. 2.

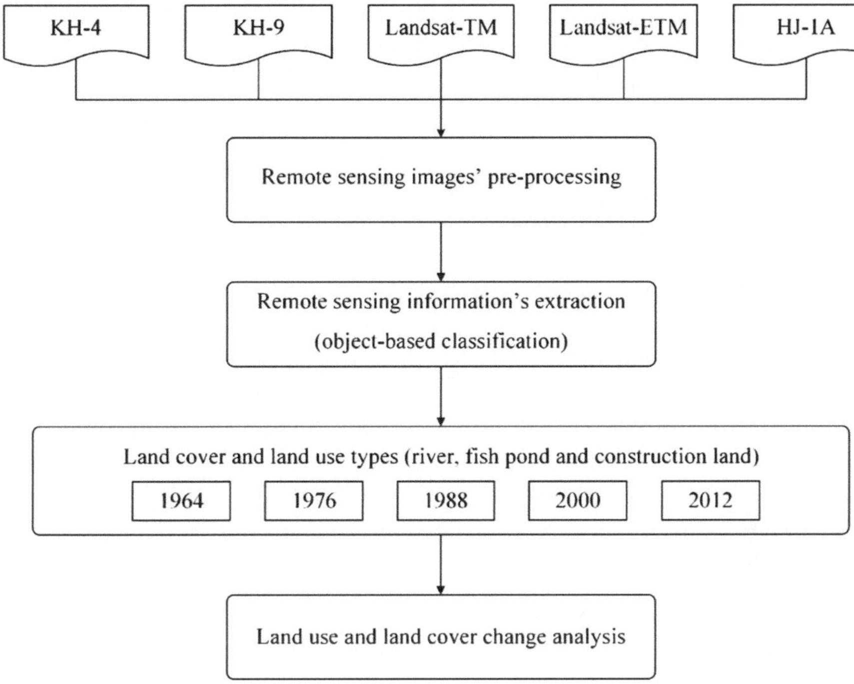


**Fig. 2** The research workflow

## *3.2 Remote Sensing Data and Data Preprocess*

We selected remote sensing data for five periods, including KH–4 (KeyHole, KH) imagery generated in 1964; and KH–9 imagery generated in 1976, from American spy satellite imagery sets, or DISP (Declassified Intelligence Satellite Photographs); Landsat TM data for 1988; Landsat ETM+ data for 1988; and the Chinese environment and disaster mitigation satellite HJ–1A data for 2012. More information for these five periods of data is shown in Table 1.

The utilization of KH–4 and KH–9 imagery not only offered distribution information for dike-pond systems in early periods in the study area, but its comparatively high resolution (KH–4: 2–3 m, KH–9: 6–9 m) also remedied the

**Table 1** Remote sensing data sets

| Date | Sensor | Resolution (m) | Band number | Cloud |
|---|---|---|---|---|
| 1964 | KH–4 | Resampled 2.5 | 1 | None |
| 1976 | KH–9 | Resampled 5.0 | 1 | None |
| 1988 | TM | 30 | 7 | None |
| 2000 | ETM | 30 | 7 | Little |
| 2012 | HJ–1A | 30 | 4 | None |

disadvantages of low resolution in early Landsat data (79 m). Meanwhile, the Landsat 7 data of 2012 had serious stripe damage in its imagery due to a satellite malfunction, thus we replaced it with HJ–1A imagery from the same time.

The preprocessing in this study mainly consisted of geographic registration, geometric correction, image mosaic and clipping, image fusion, and image enhancement. All were performed using ENVI 5.2 and ERDAS IMAGE 2014.

### 3.3 Object-Oriented Classification

Traditional methods for remote sensing information extraction include artificial visual interpretation and pixel-based computer auto-classification (supervised or unsupervised classification) (Qian and Xie 2005). The precision of visual interpretation is determined by the experience and knowledge base of the researchers in a specific study area. Drawing land use patches directly onto images ensures higher precision but requires much more time when compared with computer auto-classification. In contrast to traditional methods, object-oriented classification not only utilizes spectral information of the land surface, but it also takes advantage of geometry and structure information in the image, where the smallest unit for classification is no more than a single pixel but a more distinguishable object for further analysis and management (Du and Tian 2004). This technology overcomes many of the disadvantages generated from the single pixel unit (Guo and Pei 2010). Hence, in this study, we integrated both object-oriented technology and visual interpretation. We first employed multiresolution segmentation to remote sensing imagery after preprocessing using eCognition 9.0, a specialized software that generates and handles image objects, and then performed visual interpretation via land use distribution information and characteristics of the remote sensing image to extract required information for the five periods. We primarily extracted information applicable to the distribution of dike-pond features, permanent rivers, and developed land.

## 4 Results and Discussion

### 4.1 Analysis of the Area of Dike-Pond Patches and Dynamic Change Trends

To intuitively analyze the spatial-temporal dynamic change of the dike-pond systems from 1964 to 2012, we employed ArcGIS software to draw the spatial-temporal distribution thematic maps of dike-pond systems based on previous remote sensing data (Fig. 3).

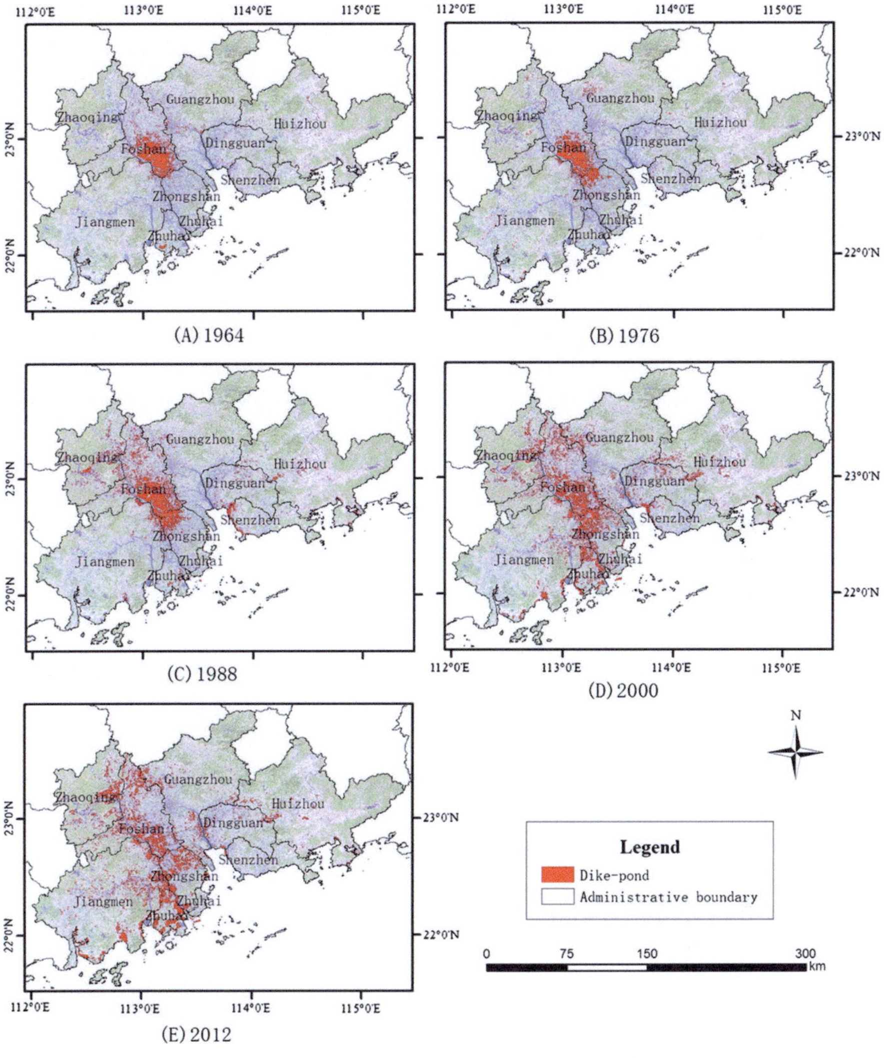


**Fig. 3** Spatial-temporal dynamic change of dike-pond systems

As shown in this figure, after nearly 50 years of development, the distribution pattern of dike-pond systems in the PRD region changed from a distribution that was mainly concentrated in Foshan City to a distribution that randomly covered the entire study area. The degree of fragmentation increased significantly, and the previous compact pattern became scattered.

In 1964, the dike-pond systems were mainly located in the Foshan Municipal District of Shunde City, southeast portions of Foshan City, and southern Nanhai City. The systems were rare in other cities. By 1976, the previous distribution had expanded to the south, and some dike-pond systems appeared to the northwest of Zhongshan City. Meanwhile some isolated dike-pond systems emerged in the western part of Guangdong City, mainly the Huadu District. Then in 1988, the previous dike-pond distribution that had been concentrated in Foshan City simultaneously spread to the northwest and southeast, and a large number of dike-pond systems appeared in Zhaoqing City. The number and size of dike-pond systems continued to grow in Zhongshan City, and eventually an aggregation developed in the northwest part of the city. In Shenzhen, some isolated dike-pond systems appeared on the coast. By 2000, dike-pond systems were already distributed all over the PRD region, and large dike-pond systems appeared in Dongguan, Huizhou, Jiangmen, and Zhuhai. Finally, in 2012, dike-pond patches covered the PRD region in a wider range but in a more fragmented pattern.

To further quantitatively discuss the dynamic change of dike-pond systems in every city, we calculated the area of dike-pond in all five periods for each city using ArcGIS software. The results are shown in Fig. 4.

For the entire research period, Foshan city always possessed the largest area of dike-pond systems in the PRD region (except in 2012), and it remained comparably stable compared to the overall fluctuations in development in other parts of the region. The area of dike-pond systems reached the lowest point in 1976

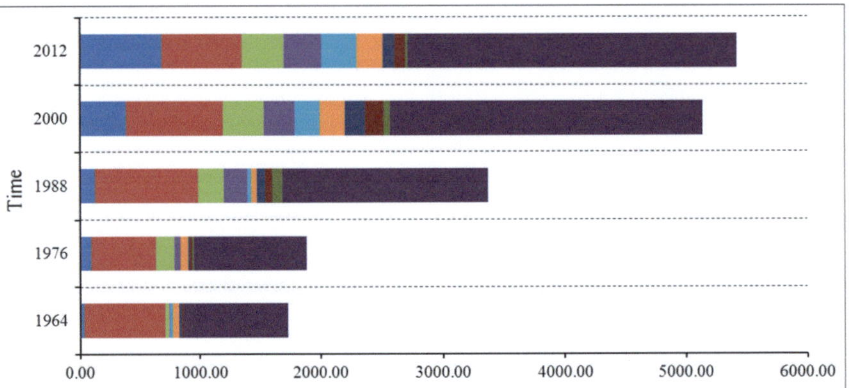


| | 1964 | 1976 | 1988 | 2000 | 2012 |
|---|---|---|---|---|---|
| Jiangmen | 35.01 | 94.41 | 125.70 | 385.54 | 683.61 |
| Foshan | 674.84 | 536.07 | 857.57 | 804.50 | 662.77 |
| Zhongshan | 32.87 | 156.57 | 217.11 | 340.11 | 353.20 |
| Zhaoqing | 5.71 | 48.40 | 189.37 | 258.62 | 304.72 |
| Zhuhai | 21.31 | 3.66 | 38.39 | 207.50 | 295.30 |
| Guangzhou | 54.46 | 59.87 | 41.17 | 203.06 | 213.44 |
| Huizhou | 24.58 | 16.37 | 73.04 | 172.44 | 94.45 |
| Dongguan | 12.84 | 16.31 | 59.35 | 144.48 | 87.33 |
| Shenzhen | 2.46 | 11.91 | 84.79 | 52.09 | 16.33 |
| Total | 864.08 | 943.57 | 1686.49 | 2568.34 | 2711.15 |

Unit: $km^2$

**Fig. 4** The change in area of dike-pond systems in every city of the PRD region from 1964 to 2012

(546.07 km$^2$) and then reached the highest point of 857.57 km$^2$ in 1988. The area of dike-ponds in Guangzhou, Jiangmen, Zhongshan, and Zhaoqing (partially) continued to increase, and reached peaks in 2012 (213.44, 683.61, 353.20, and 304.72 km$^2$, respectively). In Jiangmen, the area of dike-ponds (683.61 km$^2$) finally surpassed that of Foshan (662.77 km$^2$) in 2012. The area of dike-ponds in Zhuhai City first decreased and then increased, reaching the lowest point of 3.66 km$^2$ in 1976. The area of dike-ponds in Huizhou and Dongguan increased, decreased, and then finally increased again, in a fluctuating development pattern, which was mainly caused by the influence of multiple complex driving forces. Over the entire research period, Shenzhen City always possessed the smallest percentage of the dike-pond area in the entire PRD region, mainly due to its small area.

## 4.2 The Dynamic Transfer of Dike-Pond Land

Based on the land use classification generated from the remote sensing data for 1964–2012, we calculated the transfer matrix for dike-ponds in the PRD region (as shown in Table 2) for further analysis.

We employed the dynamic degree (DD) value of a specific land use to quantitatively describe the change in area from one period to next period in the study area (Wang and Bao 1999; Liu and Deng 2009), using the formula:

$$K = \frac{U_b - U_a}{U_a} \times \frac{1}{T} \times 100\% \qquad (1)$$

where $K$ is the DD value of a certain land use type, $U_a$ and $U_b$ are the areas of certain land uses in one period and its next period, and $T$ is the time length between the two periods. Specifically, when $T$ equals a single year, $K$ is redefined as the annual DD value of certain land use types.

**Table 2** The dynamic degree (DD) value of dike-ponds in every city from 1964 to 2012

| DD (%) | 1964–1976 | 1976–1988 | 1988–2000 | 2000–2012 |
|---|---|---|---|---|
| Guangzhou | 0.83 | −2.6 | **32.76** | 0.42 |
| Zhuhai | −6.9 | **79.07** | 36.7 | 3.52 |
| Foshan | −1.71 | **4.99** | −0.51 | −1.46 |
| Huizhou | −2.78 | **28.85** | 11.34 | −3.76 |
| Dongguan | 2.25 | **21.49** | 12.84 | −2.98 |
| Jiangmen | 14.14 | 2.76 | **17.22** | 6.44 |
| Zhaoqing | **62.3** | 24.27 | 3.04 | 1.48 |
| Zhongshan | **31.36** | 3.22 | 4.72 | 0.32 |
| Shenzhen | 32.01 | **50.99** | −3.21 | −5.72 |

The largest DD value of every city for the entire period is marked in bold

Guangzhou and Jiangmen reached their largest DD value between 1988 and 2000; Zhuhai, Foshan, Huizhou, Dongguan, and Shenzhen reached their largest DD value between 1976 and 1988; Zhaoqing and Zhongshan reached their largest DD value between 1964 and 1976; and none of the cities reached their largest DD value after 2000. The largest DD value was 79.07%, achieved by Zhuhai City from 1976 to 1988. In contrast, Shenzhen had the smallest DD value among all the values, which was −5.72 from 2000 to 2012, indicating the dike-pond system of Shenzhen significantly shrank during this period. Three cities had negative DD values from 1976 to 1964, one city had a negative DD value from 1976 to 1988, then two between 1988 and 2000, and finally four between 2000 and 2012. Politics was the main factor influencing these trends from 1964 to 1976, when the enthusiasm of farmers was impeded and dike-pond production was suppressed; consequently, the area of dike-ponds in the three cities decreased, resulting in negative DD values. Over the next 24 years, productivity was freed, and the number of cities with negative DD values decreased. After 2000, in response to urbanization and economic development, a large amount of dike-pond land was transferred into other land use types, principally land uses related to urbanization, resulting in a general decrease in dike-pond area and negative DD values for many cities.

## *4.3 Analysis of Interactions Between Dike-Pond Land and Developed Land*

Since 1976, the PRD region has been the pioneer zone for Chinese reform and the "opening-up" policy, playing an essential role in the overall planning of Chinese economic development. Due to the need for economic development, conversion of land for urbanization occurred rapidly, and this strongly affected the area of dike-pond land. To quantitatively describe the interactions between dike-pond land and developed land, we calculated the conversion matrices of the land use information between every two adjacent periods of the total four periods from 1976 to 2012, as shown in Tables 3, 4 and 5.

**Table 3** Conversion matrix of land use types from 1976 to 1988 ($km^2$)

| 1988 | 1976 | | | Area |
|---|---|---|---|---|
| | Dike-pond land | Construction land | Other land | |
| Dike-pond land | 627.18 | 16.04 | 1038.98 | 1682.20 |
| Construction land | 43.05 | 505.60 | 442.19 | 990.85 |
| Other land | 294.11 | 62.13 | 38,838.62 | 39,194.86 |
| Area | 964.34 | 583.78 | 40,319.80 | 41,867.92 |

**Table 4** Conversion matrix of land use types from 1988 to 2000 ($km^2$)

| 2000 | 1988 | | | Area |
|---|---|---|---|---|
| | Dike-pond land | Construction land | Other land | |
| Dike-pond land | 911.35 | 1.02 | 1651.52 | 2563.89 |
| Construction land | 257.96 | 925.3 | 2782.12 | 3965.38 |
| Other land | 512.98 | 64.58 | 34,760.72 | 35,338.28 |
| Area | 1682.29 | 990.9 | 39,194.36 | 41,867.55 |

**Table 5** Conversion matrix of land use types from 2000 to 2012 ($km^2$)

| 2012 | 2000 | | | Area |
|---|---|---|---|---|
| | Dike-pond land | Developed land | Other land | |
| Dike-pond land | 1357.52 | 1.37 | 1354.94 | 2713.83 |
| Construction land | 394.6 | 3826.22 | 2856.2 | 7077.02 |
| Other land | 811.77 | 138.02 | 31,126.91 | 32,076.7 |
| Area | 2563.89 | 3965.61 | 35,338.05 | 41,867.55 |

As shown in Tables 3, 4 and 5:

1. **Conversion of dike-pond land**: The area of unchanged dike-pond land continued to increase in all three time periods. From the first period (1976–1988) to the second (1988–2000), it increased by 284.17 $km^2$, and that number increased to 446.17 $km^2$ from the second period (1988–2000) to the third (2000–2012). This phenomenon is strongly relevant to the overall increase in total dike-pond area. Moreover, the area that was converted from dike-pond land to developed land increased rapidly from 43.05 $km^2$ in the first period to 257.96 $km^2$ in the second, with a difference of 214.96 $km^2$, and then continued to increase to 394.6 $km^2$ in the third period, with a difference of 136.64 $km^2$, indicating that conversion was more rapid before 2000, though the converted area was always increasing. The slowdown in conversion after 2000 indicates that the appropriation of dike-pond land caused by the expansion of urbanized and developed land decreased after 2000, whereas appropriation was much more intense due to the need for economic development before 2000.
2. **Conversion of developed/urbanized land**: Since 1988, the conversion of dike-pond land to developed land has been too small to be recognized (1.02 $km^2$ in the second period, and 1.37 $km^2$ in the third), and these values were also low in the first period from 1976 to 1988 (16.04 $km^2$). These results indicate that some of the developed land in the PRD region whose economic benefit was low was artificially converted into dike-ponds during 1976–1988, whereas this almost never happened after 1988.
3. **Conversion of other land use types**: To simplify this issue, we calculated the ratio of the area that was converted from other land use types to dike-pond land in comparison to developed land in every period. The ratio was 2.35 in the first period, then 0.59 in the second period, and finally 0.47 in the third. These results indicate that the area converted from other land use types to dike-pond land was

much larger than that converted to developed land in the first period. However, this changed after 1988 when the percentage of land converted from other land use types to developed land was much larger than that to dike-pond land. In other words, the expansion of dike-pond lands to other land use types was much stronger than that to urbanized/developed land before 1988, but that pattern reversed after that point in time.

## 5 Conclusions

We quantitatively analyzed and described the spatial-temporal distribution and evolution of the dike-pond system in the PRD region based on remotely sensed imagery for five periods. The main conclusions are summarized below:

- The area of dike-pond systems continued to increase during 1964–2012, with the rate of increase peaking during 1988–2000, and the lowest point during 1964–1976. During the entire research period, the distribution of dike-pond systems maintained trends of fragmentation and isolation.
- Due to the influence of politics from 1964 to 1976, the enthusiasm of farmers for the dike-pond systems was impeded and dike-pond production was suppressed. This led to a decrease in the area of dike-pond systems in three cities in the study area. Then during 1976–2000, with the liberation of productivity, the dike-pond industry recovered. Finally, since 2000, large areas of dike-pond systems were converted into other land use types in response to urbanization, resulting in a decrease in dike-pond system area in multiple cities.
- 1988 was a key point in time in respect to interactions between urbanized and dike-pond land. First, the appropriation of dike-pond land for urbanization was stronger before 1988 but slowed after that time. Second, before 1988, some urbanized land was transferred into dike-pond land, but this process rarely occurred after 1988. Third, the transformation of dike-pond systems to other land uses was more intense than that of urbanized land before 1988, but then it reversed when the conversion of urbanized land to other land uses surpassed that of dike-pond land.

**Acknowledgements** This work is supported by the National Science Foundation of China (Grant No. 41001291), the Science and Technology Plan Project of Guangzhou (Grant No. 2015100100 81), and the Natural Science Foundation of Guangdong (Grant No. 2016A030313261).

## References

Du F, Tian Q (2004) Object-oriented image classification analysis and evaluation. Remote Sens Technol Appl 19(1):20–23

Gongfu Z (1982) The mulberry dike-fish pond complex: a Chinese ecosystem of land-water interaction on the Pearl River Delta. Hum Ecol 10(2):191–202

Guo L, Pei Z (2010) Application of method and process of object-oriented land use-cover classification using remote sensing images. Trans CSAE 26(07):194–198
Guo C, Xu S (2011) The new visual angle and advance on researches of the dike-pond system in China. Wetland Sci 9(01):75–81
He Z, Guan L (1998) Sustainable development of mulberry dike and fish pond farming in Pearl River Delta by remote sensing technology. Acta Scientiarum Naturalum Universitie Sunyatseni 37(S2):64–69
Korn M (1996) The dike-pond concept: sustainable agriculture and nutrient recycling in China. Ambio 25(1):6–13
Li H, Luo S (2005) Reconstruction and control of modern in tensive dike-pond system in Shunde. Chin J Ecol 24(01):108–112
Liu J, Deng X (2009) The progress of spatial and temporal research methods for LUCC. Chin Sci Bull 54(21):3251–3258
Liu K, Wang S (2008) Spatial evolution analysis of dike-pond systems in Foshan City. Trop Geogr 28(06):513–517
Lo CP (1996) Environmental impact on the development of agricultural technology in China: the case of the dike-pond ('jitang') system of integrated agriculture- aquaculture in the Zhujiang Delta of China. Agric Ecosyst Environ 60(2–3):183–195
Nie C, Li H (2001) The integrated dike-pond systems: its new development and ecological problems. J Foshan Univ (Natural Science Edition) 19(01):49–53
Nie CR, Luo SM (2003) The dike-pond system in the Pearl River Delta: degradation following recent land use alterations and measures for their ecological restoration. Acta Ecol Sin 23 (9):1851–1860
Qian Q, Xie R (2005) Land cover extraction based on object oriented analysis. Remote Sens Technol Appl 20(3):338–342
Wang X, Bao Y (1999) Study on the methods of land use dynamic change research. Prog Geogr 01:83–89
Wang X, Kang Q (1997) Application of TM image in investigating changes of dike-pond land on the Pearl River Delta—samples as Shunde, Nanhai dike-pond region. Remote Sens Land Resour 3:8–14
Wang X, Xia L (2011) Spatial-temporal changes in dike-pond land in Nanhai District based remote sensing and GIS. Resour Ind 13(04):55–60
Yee AWC (1999) New developments in integrated dike-pond agriculture-aquaculture in the Zhujiang Delta, China: ecological implications. Ambio 28(6):529–533
Zhong G, Wang Z (1993) Land-water interaction of the dike-pond system. Science Press, Beijing

# Part II
# Agriculture Monitoring

# Effects of Irrigation and Nitrogen on Maize Growth and Yield Components

**Xiukang Wang and Yingying Xing**

**Abstract** The purpose of this study was to investigate the effect of irrigation and nitrogen fertilization on maize growth and yield components. Three irrigation treatments were included, 100, 150, and 200 mm, and three nitrogen levels were applied as follows: high nitrogen was 240 kg $ha^{-1}$, medium nitrogen was 180 kg $ha^{-1}$, and low nitrogen was 120 kg $ha^{-1}$. The results indicated that the interaction of nitrogen and irrigation has no significant effects on maize height, but the signal factor has a significant effect on plant height in the whole growth period. The changing trend of crop growth rate in the whole growth period was increasing first and then falling, and the CK treatment was always lower than other treatments. The maize growth rate was related to the nitrogen fertilizer level, and the positive relationship between nitrogen fertilizer level and growth rate. In single factor of fertilization, there was a negative correlation between units increased in yield and unit nitrogen. The interaction of nitrogen and irrigation has significant effects on biomass yield. The greatest yield-increasing potential was obtained in MF treatment. At the same irrigation level, the grain yield increased and had a most significant correlation relation with the harvest index. The population physiological indices of maize were increased with irrigation amount and fertilizer level, except the harvest index, and the incentive of population physiological indices in irrigation was higher than nitrogen fertilization. Therefore, MFHW treatment may be considered the most efficient for maize production in the rain-fed area of the Loess Plateau, China.

**Keywords** Maize · Irrigation · Nitrogen fertilization · Harvest index · Grain yield

X. Wang (✉) · Y. Xing
College of Life Science, Yan'an University, Yan'an 716000,
Shaanxi, People's Republic of China
e-mail: wangxiukang@126.com

X. Wang
Northwest Arid Regions of Water Conservancy Engineering State Key Laboratory,
Xi'an University of Technology, Xi'an 712100, Shaanxi, People's Republic of China



S. Pirasteh and J. Li (eds.), *Global Changes and Natural Disaster Management: Geo-information Technologies*, DOI 10.1007/978-3-319-51844-2_5

## 1 Introduction

In the last several decades, the uses of irrigation and fertilization have led to the increases in crop production and food security (Godfray et al. 2010). Irrigation and fertilization are widely used for production of food crops such as maizeand wheat, mostly because it alters the farm environment by changing the soil water contents and soil nutrients which result in the increase of soil fertility and growth environment. Therefore, it would be necessary to improve the water use efficiency and fertilizer production efficiency in arid and semiarid regions, especially for field management purposes.

Nitrogen is an essential crop nutrient, which is consumed by the crop roots throughout the growing season. Most common forms of nitrogen found in the soils are organic N, ammonium, nitrate, and gaseous nitrogen. Nitrate leaching is the main loss of nitrogen fertilizer which will result in its accumulation in the deeper soil layers, and will make the groundwaterpolluted (Wang et al. 2015). Nitrate leaching potential depends on irrigation methods, soil properties, crops and crop rotation, management practices and climatic parameters. This necessitates the development of appropriate water and fertilizer application strategies so as to maximize their application efficiency and minimize fertilizer losses through leaching (Ajdary et al. 2007). The development of fertigation technology has provided a new technique for the high-efficient and high-yield crop production. In particular, fertigation technology has proven to be one of the most effective methods to increase water use efficiency (WUE), grain yieldand crop quality in dry farming agricultural areas (Singandhupe et al. 2003). Studies have indicated that fertigation technology is conducive to crop growth by decreasing the soil moist and reducing soil surface evaporation in dryland agriculture (Badr et al. 2010). Fertigation technology also has the benefit of improving the absorption of crops on soil nutrient, including protection of the stability of fertilizer on root zone. Studies have also demonstrated that the benefits of deficit drip irrigation of crops are the results of water saving 33%, improving irrigation WUE by 42%, and improving the maize yield and quality (Kaman et al. 2011). The absolute growth rate, spike weight, spike length, 100-grain weight, and yield were bigger in fertigation technology than furrow irrigation.

Previous studies have revealed that the maximizing total production of maize under irrigation and nitrogen management (Wang and Xing 2016). However, few investigations have investigated on how irrigation and nitrogen presently affects maize yield in Northwest China (Sun et al. 2011), particularly the modeling of water and nitrogen coupling were not taken into consideration. Moreover, the study on coupling effect of irrigation and fertilizer is an important part of the semiarid region. Until recently, there is also need for the coupling effect of irrigation and fertilizer on maize yield, and there are fewer results on effects of the interaction. In this paper, we would examine how irrigation and nitrogen fertilizationaffectsmaizegrowth, yield, and setup production models.

# 2 Materials and Methods

## 2.1 Experimental Sites

Field experiments were conducted in the Loess Plateau area at the Ansai Experimental Station (36° 52′N, 108° 10′E and altitude 1320 m) in Shanxi Province, Northwest China in 2014. The location has a semiarid and warm temperate climate with a mean annual temperature of 9.1 °C and an annual accumulated temperature (>0 °C) of 3150 °C. The mean annual precipitation is 580 mm, and annual pan evaporation-based potential evaporation is 1500 mm. The average annual sunshine duration is 2230 h, and there are over 171 frost-free days. The groundwatertable is approximately 50–80 m below the surface. The permanent wilting point was 22.3%. As shown in Fig. 1, the maximum temperature is very high, and the relative humidity is very low during the growing season.

## 2.2 Experimental Design

Maize (Zea mays, L. Liyu 18) seeds were sown on May 6 and harvested on September 21, 2014 (days of the year: 138). The experiment used randomly assigned field plots with three replicates per treatment in two years. There were three nitrogen fertilization and three irrigationtreatments. The three nitrogen (urea) fertilization was applied as follows: HN (240 kg/ha), MN (180 kg/ha), and ZN (120 kg/ha), and the three irrigation treatments were HW (200 mm), MW (150 mm), and LW (100 mm).

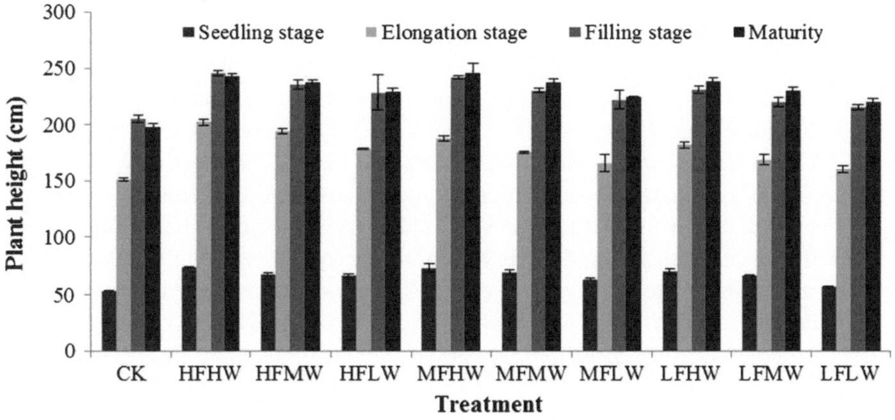


**Fig. 1** Effects of different nitrogen and irrigation on plant height

**Table 1** Different nitrogen and irrigation amount and time in maize growing season, the maize growing season was divided into four periods and the irrigation time in different periods according to the rainfall time

| Fertilization | Irrigation | Seedling | Elongation | Filling | Maturing | Irrigation amount |
|---|---|---|---|---|---|---|
| CK | | 0 | 0 | 0 | 0 | 0 |
| HF | HW | 50 | 50 | 50 | 50 | 200 |
| | MW | 37.5 | 37.5 | 37.5 | 37.5 | 150 |
| | LW | 25 | 25 | 25 | 25 | 100 |
| MF | HW | 50 | 50 | 50 | 50 | 200 |
| | MW | 37.5 | 37.5 | 37.5 | 37.5 | 150 |
| | LW | 25 | 25 | 25 | 25 | 100 |
| LF | HW | 50 | 50 | 50 | 50 | 200 |
| | MW | 37.5 | 37.5 | 37.5 | 37.5 | 150 |
| | LW | 25 | 25 | 25 | 25 | 100 |

In this study, three check treatments were without irrigation, and the other treatments were examined as Table 1. Irrigation began when 60% of the surface field capacity (0–60 cm) was depleted by using the oven method, which generally occurred at the beginning of May in both years. The size of each experimental plot area was 10 $m^2$ (6 m × 3 m = 18 $m^2$). Planting density was 47,220 ha/plant. Each plot at planting time, spacing of 35 cm, 60 cm row spacing planting, planting depth of soil generally plummeted to the original seedling roots.

## 2.3 Measurements

Plant heights were determined by measuring the heights of two plants in one plot, which were measured from the ground level to the tips of the tassels at physiological maturity, from which the average plant height was determined. Crop growth rate is calculated as (CGR) = $(M2 - M1)/(T2 - T1)$, where CGR is crop growth rate; $T1$, $T2$ are days after sowing. $M1$ and $M2$ of $T1$ and $T2$, respectively, determine the amount of dry matter accumulation. In the harvest field experiments, the maize yield was calculated using the plot yield calculation method. The mature period randomly selected ten strains in every community, roots were dug up, the process was repeated three times; after aboveground harvest, straw, root, and grain were separated and air dried after weighing; the quality of dry matter, grain yield, and yield components were measured.Irrigationwater use efficiency (kg/$m^3$) was calculated by dividing the grain yield (kg/ha) by irrigation amount ($m^3$/ha). The irrigation water productivity was the ratio of the biological yield and irrigation water.

### 2.4 *Statistical Analysis*

An analysis of variance was conducted on the plant height, leaf area, root dry weight, shoot dry weight, and grain yield with SAS 9.2 (SAS Institute Ltd., North Carolina, USA). Duncan's multiple range tests were used for paired mean comparisons at a 0.05 probability level.

## 3 Results and Discussion

### 3.1 *Plant Height*

Plant height is an important index of maize growth, and the effects of different nitrogen and irrigation on plant height in whole growing season are shown in Fig. 1. In the whole growth period, plant growth was fast from seedling stage to elongation stage; in the filling stage, plant height growth slowed and mature plant height declined slightly. The CK treatment was significantly lower than other treatments, the interaction of nitrogen and irrigation has no significant effects on plant height, but the signal factor has a significant effect on plant height in the whole growth period. The plant height had a positive correlation with irrigation and fertilizer amount.

The highest plant height was 73.30 cm in HFHW treatment in seedling stages, and the average of plant height in HF was only 0.63 cm higher than MF treatment. However, the average of plant height in HW was 4.83 cm higher than MW treatment. The results might indicate that the plant height was sensitively in water than fertilizer. There was a significant difference between plant height in HF and LF treatment at the same irrigationlevel in elongation stage. The elongation stage is the fastest growth period of maize plant growth, and the average growth speed of plant height in HW, MW, and LW treatments were 4.94, 4.43, and 4.68 cm day$^{-1}$, respectively. There was no significant influence on plant height in irrigation and fertilizer treatment in filling stage, and the highest plant height was found in MFHW treatment, and the speed of plant growth was changed to slow. The average increase speed of plant height was 2.79, 3.44, and 3.23 cm day$^{-1}$ in HF, MF and LF treatments, respectively. The results indicated that filling stage of plant growth rate is less than the rate of increase in elongation stage. In mature period, the plant height reduced except the HW treatment which may relate that the HW processing in the mature period of plant height has a certain role in promoting plant height increase. The research reported that maize height was not notable with the addition of nitrogen only improving its effect by about 3% (Midega et al. 2013). In most cases, the height of maize was not significantly different between the nitrogen levels and irrigation frequencies (Hokam et al. 2011).

**Table 2** Models of different irrigation and fertilization on maize growth rate; *y* is the dry weight accumulation, *x* is days after sowing, and $R^2$ is the regression fitting the degree of relationship as the model

| Treatments | Equation | $R^2$ |
|---|---|---|
| CK | $y = -0.0063x^2 + 0.947x - 19.925$ | 0.8681 |
| HFHW | $y = -0.0077x^2 + 1.37x - 30.466$ | 0.8873 |
| HFMW | $y = -0.0089x^2 + 1.3723x - 29.485$ | 0.8644 |
| HFLW | $y = -0.0103x^2 + 1.4195x - 29.651$ | 0.8461 |
| MFHW | $y = -0.0073x^2 + 1.2718x - 27.796$ | 0.9141 |
| MFMW | $y = -0.0073x^2 + 1.1758x - 25.161$ | 0.9074 |
| MFLW | $y = -0.0095x^2 + 1.3396x - 28.093$ | 0.8410 |
| LFHW | $y = -0.0082x^2 + 1.2729x - 27.25$ | 0.8755 |
| LFMW | $y = -0.0092x^2 + 1.3434x - 28.588$ | 0.8260 |
| LFLW | $y = -0.01x^2 + 1.3892x - 29.399$ | 0.7748 |

## 3.2 Crop Growth Rate

The field of maize growth rate with time change model under different irrigation and fertilization treatments as shown in Table 2, the relationship between maize-growth rate and growth time was consistent with polynomial relation and the better fitting coefficient of different treatments. The best fitting was MFHW treatment in this experimental, and the fitting coefficient was 0.9141, and the fitting coefficient was above 0.82 except LFLW treatment. The changing trend of crop growth rate in the whole growth period was increasing first and then falling, and the CK treatment was always lower than other treatments.

In seedling stage, the influence of different irrigationand fertilization on maize growth rate was not significant, and the growth rate was ranged from 1.293 to 1.534 g $m^{-2}$ $day^{-1}$ (Fig. 2). The growth rate of maize increased rapidly at elongation stage, and the growth rate was ranged from 8.538 to 13.008 g $m^{-2}$ $day^{-1}$. The growth rate continued to increase in the filling stage, the HW, MW, and LW treatment of maize growth rate in high fertilizer levels were higher 75.76, 72.21 and 47.89% than CK, respectively. In the mature period, the growth rate was declined, but the rate of decline was negatively related to the fertilizer rate. The result indicated that the starting point of growth rate decline was related to the fertilizer rate, the higher of fertilizer amount, and the later decline point appeared of growth rate.

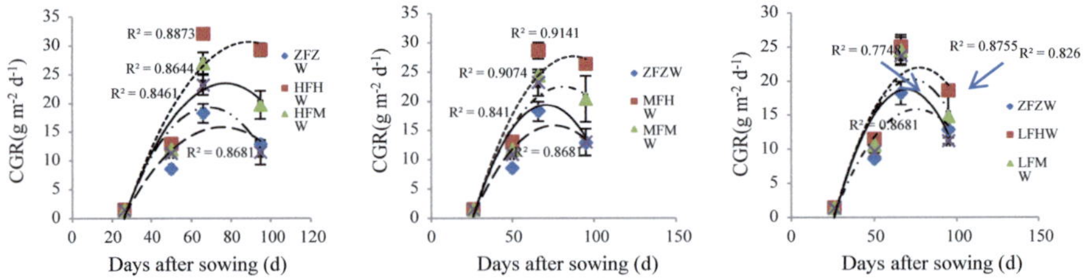


**Fig. 2** The relationship between crop growth rate and days after sowing, CGR was crop growth rate

## 3.3 Yield and Its Components

The effect of different irrigation and fertilization on maize yield and its components are shown in Table 3. The spike length, spike weight, grain weight per spike, the 100-grain weight and yield increased with the increase of fertilizer application and irrigation amount, the nitrogen level very significantly affected the grain weight per spike and yield, and the irrigation level significantly affected the spike length and weight. The highest spike length was 17.34 in HFHW treatment which was 1.218 times higher than CK treatment, and the spike length was ranged from 14.24 to 17.34 cm in different irrigation and fertilization levels. Spike grain weight directly affects crop yields which were distributed in the range from 145.17 to 236.15 g $plant^{-1}$, and the highest spike weight was 236.15 g $plant^{-1}$, increased 62.7% than CK treatment. There was no significant difference between HW and MW treatment, but the HW and LW grain weight per spike significantly differed. HFMW, MFHW, and MFMW treatment than HFHW spike grain weight was reduced by 3.58, 1.37, and 5.20%, respectively, the results show that HFHW production efficiency was low. The 100-grain weight is another evaluation index of maize yield, which is the direct influence of the size and plumpness of maize seeds. The maximum of 100-grain weight was 33.9 g in HFHW treatment, which was significantly higher than the LFLW and CK treatment, but no significant difference with other treatment.

**Table 3** Different irrigation and fertilization on maize yield and its components

| Fertilization | Irrigation | Spike length (cm) | Spike weight (g $plant^{-1}$) | Grain weight per spike (g $plant^{-1}$) | 100-grain weight (g) | Yield (kg $ha^{-1}$) |
|---|---|---|---|---|---|---|
| CK | | 14.24c | 145.17d | 116.58d | 25.84d | 4599.09d |
| HF | HW | 17.34a | 236.15a | 196.39a | 33.90a | 7923.02a |
| | MW | 16.83ab | 228.57a | 189.36ab | 32.32abc | 7600.50ab |
| | LW | 16.34b | 207.68b | 172.38bc | 31.68abc | 6918.94bc |
| MF | HW | 17.21a | 230.39a | 193.70a | 33.33ab | 7774.90a |
| | MW | 16.75ab | 218.90ab | 186.18ab | 32.14abc | 7473.06ab |
| | LW | 16.30b | 202.42bc | 175.52bc | 31.41bc | 7044.98bc |
| LF | HW | 16.31b | 202.33bc | 187.60ab | 32.63abc | 7529.85ab |
| | MW | 16.09b | 189.71c | 164.20c | 31.64abc | 6590.60c |
| | LW | 16.14b | 185.50c | 159.10c | 30.63c | 6386.10c |
| Nitrogen level | | NS | NS | ** | NS | ** |
| Irrigation level | | ** | * | NS | NS | NS |
| Nitrogen × irrigation | | NS | NS | * | NS | * |

Values followed by the same uppercase letter in the same row are not significantly different according to Duncan's multiple range test ($P < 0.05$)

"**" means $0.001 < p < 0.01$, "*" means $0.01 < p < 0.05$ and "NS" means $p > 0.05$

The maize yield declined when irrigation and fertilizer reduced, the influence of fertilizer application on yield was very significant, but the influence of irrigation water for production was not significant, and the interaction of irrigation and fertilizer significantly affected on maize yield. The highest maize yield was 7923.02 kg ha$^{-1}$ in HFHW treatment, which was 41.95% higher than CK treatment. The results might be maize did not fully utilize seasonally available water due to percolation below the root zone or water left in the ground at seedling and mature period (Grassini et al. 2009). The other reasons might include the fastest growing stage of maize that had a rapid development during the elongation stage and needed a lot of water and nutrients (Suyker et al. 2005). The highest root and shoot dry weight were obtained in treatment HN4, which might be explained by the high amount of irrigation and nitrogen fertilizationcan meet the increases in air temperature and solar radiation (Romano et al. 2011). In same high irrigation amount, the average maize yield of HF, MF, and LF has increased by 13.85, 17.64, and 24.42 kg per kilogram of nitrogen fertilizer. Results show that under the single factor of fertilization, there was a negative correlation between units increased in yield and unit nitrogen.

## 3.4 Water Use Efficiency and Harvest Index

Effects of different treatments on harvest index, water production, water use efficiency, and biomass yield of maizeare shown in Table 4. The photosynthetic product quantity was directly on the accumulation of biomass yield, which was also the foundation of into maize yield. The biomass yield was increased with the amount of irrigationand fertilizer. The CK treatment was significantly lower than others, the biomass yield of HFHW treatment was 1.777 times than CK treatment. The results indicated that nitrogen fertilization at 175 kg ha$^{-1}$ (1050 mg/pot) has

**Table 4** Effects of different treatments on biomass yield (kg ha$^{-1}$), harvest index (HI), water production (WP, kg m$^{-3}$), and water use efficiency (WUE, kg m$^{-3}$) of maize

| Treatments | Irrigation (mm) | Biomass yield (kg ha$^{-1}$) | WP (kg m$^{-3}$) | WUE (kg m$^{-3}$) | HI |
|---|---|---|---|---|---|
| CK | 0 | 11,646.25 | – | – | 0.395 |
| HFHW | 200 | 20,698.12 | 8.624 | 3.962 | 0.383 |
| HFMW | 150 | 16,396.22 | 9.109 | 5.067 | 0.464 |
| HFLW | 100 | 14,256.22 | 10.560 | 6.919 | 0.485 |
| MFHW | 200 | 19,078.87 | 7.950 | 3.887 | 0.408 |
| MFMW | 150 | 15,980.17 | 8.878 | 4.982 | 0.468 |
| MFLW | 100 | 14,678.95 | 10.873 | 7.045 | 0.480 |
| LFHW | 200 | 15,262.15 | 6.359 | 3.765 | 0.493 |
| LFMW | 150 | 13,758.50 | 7.644 | 4.394 | 0.479 |
| LFLW | 100 | 13,556.18 | 10.042 | 6.386 | 0.471 |

significant increase in plant height, root weight, and biomass in response to inoculation (Gajri et al. 1997; Shaharoona et al. 2006). The optimal growth of maize in semiarid regionsis similar to that in Turkey under two days irrigation frequency and 100% ET water application (Oktem et al. 2003). The same result was observed that the highest biomass yield with 200 mm irrigation amount in the same nitrogen fertilization. The results showed that obvious influence of nitrogen and irrigation on biomass yield, and the interaction of nitrogen and irrigation have significant effects on biomass yield.

The irrigation water productivity was increased with nitrogen application at the same irrigationtreatment, and the water productivity had a negative correlation with irrigation amount under the same fertilizer levels. The results indicated that nitrogen could improve the water production and water use efficiency. The highest water productivity was 10.873 kg $m^{-3}$ in MFLW treatment, and the water use efficiency was 7.054 kg $m^{-3}$. The relationship of group source and library was harvest index, which may also reflect the Groups of yield components and the accumulation, distribution of the photosynthetic product. There was significant effect of nitrogen and irrigation amount on harvest index, and the harvest index ranged from 0.383 to 0.493. In HF and MF treatment, the harvest index had a negative correlation with irrigation amount, and there was a positive correlation between harvest index and irrigation at the LF levels. Results show that improving the harvest index could achieve strong and sink and source of double effect, finally can achieve the purpose of increasing production.

## *3.5 Population Physiological Indices*

Effects of different nitrogen and fertilizer on population physiological indices of maizeare shown in Fig. 3. At the same levels of fertilizer, the maize yield was positively correlated with irrigation amount. The highest grain yield was obtained in HFHW treatment, and the greatest yield increasing potential was obtained at MFML treatment. There was no significant difference in biomass yield in irrigation treatments, and the highest harvest index was obtained in LFHW treatment. In high irrigation levels, the maize yield, biomass yield, and crop growth rate in HF were 14.51, 45.18, and 60.08% higher than CK treatments, respectively, but the harvest indexwas reduced 21.13%. In high fertilizer levels, the maizeyield, biomass yield, and crop growth rate were increased by 5.22, 35.62, and 34.14% in HF than CK treatments, respectively, but the harvest index was reduced 22.41%. The results show that the population physiological indices of maize were increased with irrigationamount and fertilizer level, except the harvest index. However, the incentive of population physiological indices in irrigation was higher than nitrogen fertilization.

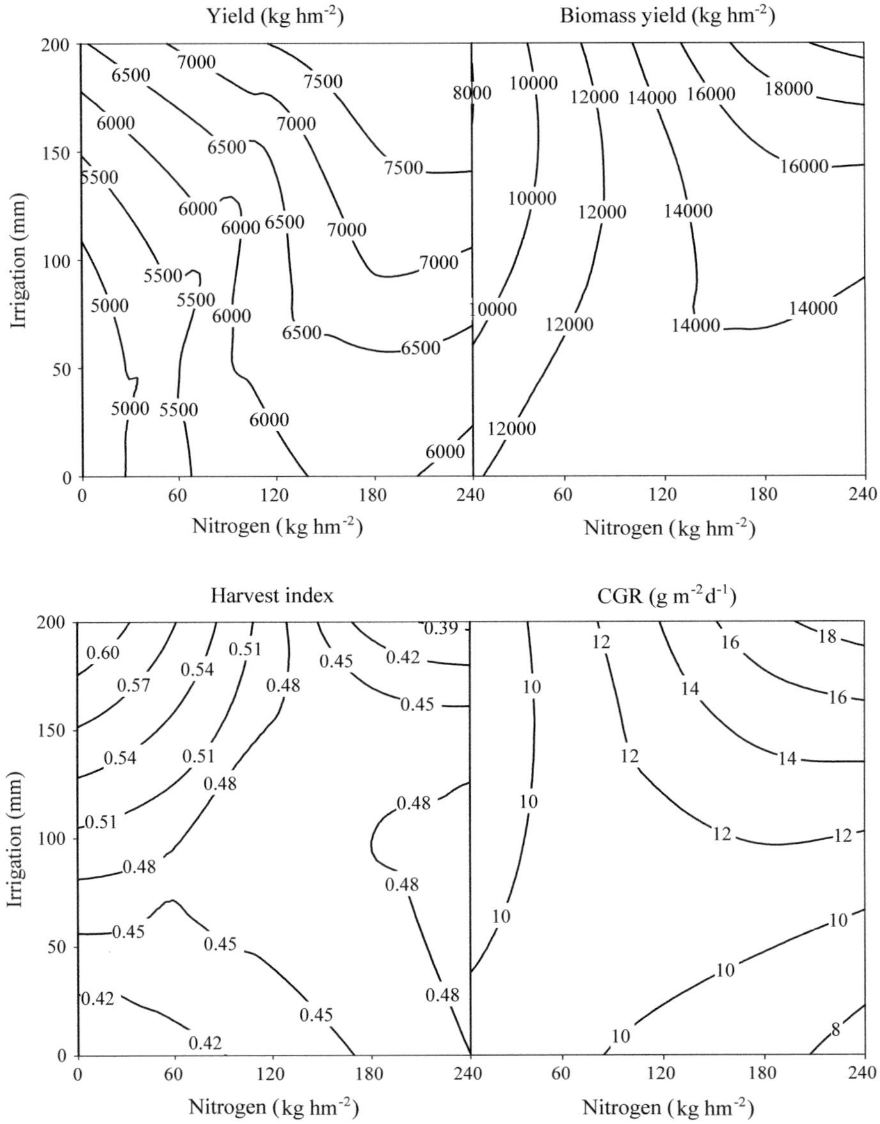


**Fig. 3** Effects of different nitrogen and fertilizer on population physiological indices of maize

# 4 Conclusions

The highest maize yield was 7923.02 kg ha$^{-1}$ in HFHW treatment, which was 41.95% higher than CK treatment. In filling stage, the growth rate of HW and MW treatment was higher 75.76 and 72.21% than CK. The highest water productivity was 10.873 kg $m^{-3}$ in MFLW treatment, and the water use efficiency was 7.054 kg $m^{-3}$. For these results and more, MFHW treatment may be considered the most efficient for maize production in the rain-fed area of the Loess Plateau, China.

**Acknowledgements** The authors are thankful to the editors and anonymous reviewers for their valuable comments on this paper. This research was supported by specialized research fund for the doctoral program of National Natural Science Foundation (51669034), Yan'an University (YDBK2015-05, YDBK2015-09), Special research in Shaanxi province department of education (16JK1853) and Shaanxi Province High-level University Construction Special Fund Projects of Ecology (2012SXTC03).

## References

Ajdary K, Singh DK, Singh AK, Khanna M (2007) Modelling of nitrogen leaching from experimental onion field under drip fertigation. Agric Water Manage 89:15–28

Badr MA, Abou Hussein SD, El-Tohamy WA, Gruda N (2010) Nutrient uptake and yield of tomato under various methods of fertilizer application and levels of fertigation in arid lands. Gesunde Pflanz 62:11–19

Gajri PR, Singh J, Arora VK, Gill BS (1997) Tillage responses of wheat in relation to irrigation regimes and nitrogen rates on an alluvial sand in a semi-arid subtropical climate. Soil Till Res 42:33–46

Godfray HC, Beddington JR, Crute IR, Haddad L, Lawrence D, Muir JF, Pretty J, Robinson S, Thomas SM, Toulmin C (2010) Food security: the challenge of feeding 9 billion people. Science 327:812–818

Grassini P, Yang H, Cassman KG (2009) Limits to maize productivity in Western Corn-Belt: a simulation analysis for fully irrigated and rainfed conditions. Agr Forest Meteoro 149: 1254–1265

Hokam EM, El-Hendawy SE, Schmidhalter U (2011) Drip irrigation frequency: the effects and their interaction with nitrogen fertilization on maize growth and nitrogen use efficiency under arid conditions. J Agron Crop Sci 197:186–201

Kaman H, Kirda C, Sesveren S (2011) Genotypic differences of maize in grain yield response to deficit irrigation. Agric Water Manage 98:801–807

Midega CAO, Pittchar J, Salifu D, Pickett JA, Khan ZR (2013) Effects of mulching, N-fertilization and intercropping with *Desmodium uncinatum* on *Striga hermonthica* infestation in maize. Crop Prot 44:44–49

Oktem A, Simsek M, Oktem AG (2003) Deficit irrigation effects on sweet corn (Zea mays saccharata Sturt) with drip irrigation system in a semi-arid region part I. Water-yield relationship. Agric Water Manage 61:63–74

Romano G, Zia S, Spreer W, Sanchez C, Cairns J, Araus JL, Müller J (2011) Use of thermography for high throughput phenotyping of tropical maize adaptation in water stress. Comput Electron Agr 79:67–74

Shaharoona B, Arshad M, Zahir Z, Khalid A (2006) Performance of Pseudomonas spp. containing ACC-deaminase for improving growth and yield of maize (Zea mays L.) in the presence of nitrogenous fertilizer. Soil Biol Biochem 38:2971–2975

Singandhupe RB, Rao GGSN, Patil NG, Brahmanand PS (2003) Fertigation studies and irrigation scheduling in drip irrigation system in tomato crop (*Lycopersicon esculentum* L.). Eur J Agron 19:327–340

Sun Y, Yin J, Cao H, Li C, Kang L, Ge F (2011) Elevated CO2 influences nematode-induced defense responses of tomato genotypes differing in the JA pathway. PLoS ONE 6:e19751

Suyker AE, Verma SB, Burba GG, Arkebauer TJ (2005) Gross primary production and ecosystem respiration of irrigated maize and irrigated soybean during a growing season. Agric Forest Meteoro 131:180–190

Wang X, Li Z, Xing Y (2015) Effects of mulching and nitrogen on soil temperature, water content, nitrate-N content and maize yield in the Loess Plateau of China. Agric Water Manage 161:53–64

Wang X, Xing Y (2016) Effects of mulching and nitrogen on soil nitrate-N distribution, leaching and nitrogen use efficiency of maize (Zea mays L.). PLoS ONE 11(8):e0161612

# A Review of the Effects of Drought on the Grain Yield in the Vays, Mollasani, and Salamat Regions of the Khuzestan Province

**Peyman Molaali and Omosalameh Babaee**

**Abstract** The issue of drought and its effect on crops especially wheat, which is considered to be a strategic crop is of the utmost importance. In this research, the effects of drought on the yield of irrigated wheat in the Northeast irrigation network of the Khuzestan province (i.e., the Vays, Mollasani, and Salamat regions) over the years spanning 2006–2010 has been taken into account. To carry out the research, an SPI index (Standardized Precipitation Index) using the rainfall collection data of nine metrological stations for the specified drought years was used. To check the validity of the results obtained from the above index mentioned, the data was cross-referenced with ETM+ multispectral images obtained from the Landsat 7 satellite. One of the main difficulties associated with the research was the blurred ETM+ images, which is due to an SLC-off error during the period under study. However, using additional images (images taken before or after the main image) and the application of a mathematical algorithm using an attached module to the software ENVI (gap fill) to modify the band gaps in the blurred image. This resulted in recovering most of the information in the images after which the NDVI Index (Normalized Difference Vegetation Index) was applied. The images during the study period have determined the crop concentration in the area. Since the yield of irrigated wheat was the objective, another problem faced by the researchers was the fact that the low resolution of the multispectral images of the Landsat satellite (with a resolution of 30 m) did not allow for determining the boundary between the various plots or types of crops. To rectify the problem, agricultural cadastral images and descriptive information (such as ownership documents) of the region were used. By superimposing the data related to the boundaries of the farmland and the images related to the Normalized Difference Vegetation Index with the information

P. Molaali (✉)
Department of GIS and Cadaster, Khuzestan Water and Power Authority, Ahvaz, Khuzestan Province, Iran
e-mail: paliz1972@gmail.com

O. Babaee
Department of Geography, University of Payame Noor, Tehran, Tehran Province, Iran
e-mail: o.babaee43@gmail.com

 
S. Pirasteh and J. Li (eds.), *Global Changes and Natural Disaster Management: Geo-information Technologies*, DOI 10.1007/978-3-319-51844-2_6

related to the plots of land where wheat was cultivated during the study period, the amount of wheat yield during the drought years was determined. The obtained results were compared with the analyzed data obtained from GIS imagery and its juxtaposition using SPI indices showed a correlation between the occurrence of drought in the study period and a decrease in the amount of irrigated wheat yield. With due consideration to the repetition and duration of the drought in the future, it is necessary that drought preventive management strategies be applied in the agricultural sector and in addition to the optimal use of water in the irrigation network, modern irrigation techniques be applied in the farms. What more, it is proposed that existing agricultural patterns be changed and that a sturdier and more resistant variety be used during low water periods.

**Keywords** SPI index · Geographic information system (GIS) · Remote sensing (RS) · Drought · Wheat · Yield · Vays · Mollasani · Salamat

## 1 Introduction

Nowadays, remote sensing has become a useful technique to extract data for analysis and analysis of satellite images managed by the GIS in various fields of land, use of technology in environmental crises such as drought management to estimate damages, especially in agriculture and its related products, as well as to predict and manage performance before and during the crisis, including the cases studied in this emerging science.

Drought disaster is not an end in itself, but its impact on people and the environment determine whether it was catastrophic. So, the key point in understanding drought, understanding natural, social, and economic impact that result due to water shortage, damage to agricultural production , or the lack of it is its social and economic impact.

Wheat is one of the strategic crops in areas where drought occurs in its performance (production tonnage per hectare) volatility is high. The role of strategic crops such as wheat in the food security of communities is very important. The concept of long-term food security strategy to implement measures to provide access to food and nutritional requirements for all communities of the world. The volume of nutrients needed, in terms of quantity and quality in relation to the population of many countries is calculated. In fact, guarantee food security and meet human nutritional needs. In addition, food security is a necessary condition for achieving political security, social, and global economy.

Iran, in addition to the position of the arid and semi-arid on earth that is the world of the negative impact of climate change caused by increasing human destructive activities that lead to global warming, has become more vulnerable and is not safe. Such that the periods of drought in countries that lack proper planning in crisis management under drought, which has resulted in the production of strategic crops, such as wheat in different areas of fertile land to produce a stable, despite the not (that is, be self-sufficient). This has caused our country as one of the world's leading importer of this product is strategic in terms of economic damage, so always be imported into the country.

Continuing drought is one of the most serious dangers and negative consequences of climate change in different regions of the globe, including the Middle East that Iran is also part of this strategic region. So that these phenomena on water resources which are included in the Karoon River, Karkhe River, Dez River, Zohreh, and Jarahi River in Khuzestan province (South-West of Iran), and about 33% of the country's surface water resources are devoted to the impact the basin is put Karun in terms of economic, social and political decision-makers and researchers has been the focus of attention.

In the irrigation network of irrigated agriculture, drought direct impact on agricultural products. (Agricultural drought) the sensitivity of agricultural products and need more water, more losses due to the drought that product comes in So that product quality is poor may even lead to the destruction of the product, in this case, is causing serious damage to agriculture and farmers. So that it causes confusion among farmers is always the question is raised them as a producer of those circumstances and in case of when is the continuation of this situation will how to deal with it?

If the drought continues and every year suffer large losses to the agriculture sector is a farmer and the farmer is the existence, and it could be possible in spite of internal and your family profession, forced to completely abandon agriculture and urban consumers join the crowd (migrate to cities).

# 2 Materials and Methods

## *2.1 Characteristics of the Study Area*

The study area is part of Karun River catchment area. This basin is located in the central Zagros Mountains that busted many branches on Kohgiluyeh and Boyer-Ahmad province. This is by crossing the mountainous regions in the north of the province, enters the plains and then crossed the Khuzestan plain and Shoat-Gargar two branches (Small River) divided.

The two branches come together not to Ahwaz, the irrigation network Veys, Mollasani, and Salamat (the study area) between one of two branch flow path on the left side, and Ahvaz is located. VAYS, Mollasani, Salamat irrigation network, with an area of approximately 13,500 ha (135 km) northeast of the provincial capital of Ahvaz in Khuzestan province located 15 km away from construction operations began in 1382. Since 1365 it has been exploited. Due to the height difference of the Karun River network, water supply network by a central pumping station and then through the main channel, the main artery pumping water from irrigation network is central to occur. Water extracted by five sets of electro-pump that are each with a capacity of 2.5 $m^3$ of Karun River water pumped into the network (Fig. 1).

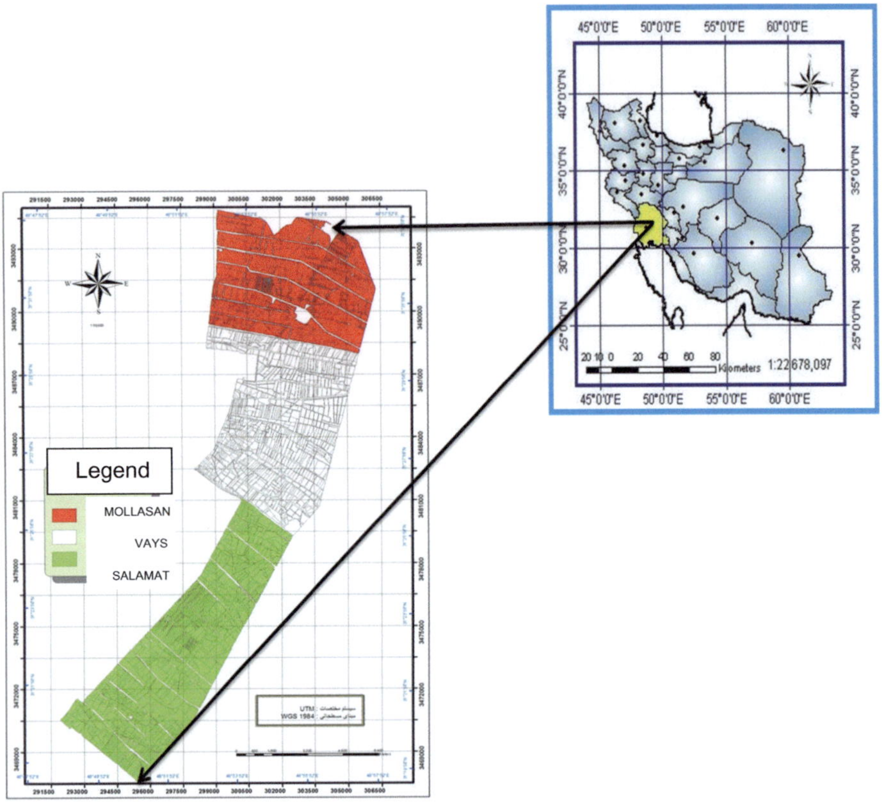


**Fig. 1** Location of the study area

## *2.2 The General Steps of the Project*

- The use of SPI index to evaluate the phenomena of drought in the region.
- The use of satellite images to extract useful data.
- The use of GIS for spatial data analysis.

### 2.2.1 The Use of SPI Index to Evaluate the Phenomena of Drought in the Region

Standardized Precipitation Index (SPI)

Index SPI (Standardized Precipitation Index), an index that depends on the probability of precipitation for any time scale, it can be calculated for different time scales and can be used for drought monitoring, and early warning helps assess its severity (Hejazizadeh 1389).

The method by McKee, Doesken, Qulest, members of Colorado climate in 1993, according to the different effects of the shortage of rainfall on groundwater reserves and surface water resources, soil moisture, and streamflow are presented.

SPI index is calculated by the following equation:

$$\mathrm{SPI} = \frac{Pi - \bar{p}}{S} \tag{1}$$

$P$ amount of precipitation during the period in question.
$\bar{p}$ long-term average rainfall for the period in question.
$S$ the standard deviation of rainfall.

SPI index values thus long-term precipitation data for a period to obtain, from a normal distribution with mean zero and standard deviation follow that one. The numerical result leads to wet and dry climates can be studied in the same way. Due to the properties of SPI, the indicators around the world to dry spells and drought monitoring is used. Including by Yamuh and his colleagues have shown, the use of SPI results to determine the effects of climate on plant performance is very helpful.

Standard Precipitation Index, is a powerful tool in the analysis of rainfall data. The aim SPI, rainfall is assigned a numerical value to areas that can be compared completely different climates. For example, what in tropics, drought is causing the same amount of rainfall as heavy rainfall in desert areas will be considered.

Among the advantages of SPI are as follows:

- Simple SPI
- Versatility SPI, to monitor the situation from the perspective of meteorological and hydrological drought
- SPI normal distribution
- Lack of dependence on soil moisture and the possibility of use in all months of the year
- SPI flexibility to scale and different time

Index SPI, calculated in the range of +2 or more for acute wet, and 2 or less to be graded severe drought. In this classification, for a period of drought appears when the index SPI, takes less than 1.

If the drought ends, it can be said that this index is a positive value. Classification based on the calculation methods listed in the index SPI, the study area in terms of drought and trends can be examined.

SPI in two forms: short-term (9, 6, 3, 1 months) and long-term (72, 28, 24, 12 months) is calculated. At any time scale, SPI, rainfall precipitation that particular period with the same period for all years for which information is recorded, provides. SPI short-term, reflecting the soil moisture conditions and rainfall estimates reveal seasonal. In this model, there are more changes and SPI index showed that the short-term, more sensitive to changes in humidity conditions and, as n becomes longer, precipitation new moon, have less effect on the overall index annually, and respond slowly. Short-term time scale, greater volatility than long-term type and is

**Table 1** Drought classification based on the SPI

| Drought | The SPI values |
|---|---|
| Mild drought | 0 to −0.99 |
| Moderate drought | −1 to −1.49 |
| Severe drought | −1.50 to −0.199 |
| Very severe drought | −2.0 |

**Table 2** SPI index value and the cumulative probability associated with it

| SPI | Cumulative probability |
|---|---|
| −3 | 0.0014 |
| −2.5 | 0.0062 |
| −2 | 0.0228 |
| −1.5 | 0.0668 |
| −1 | 0.1587 |
| −0.5 | 0.3085 |
| 0 | 0.5 |
| 0.5 | 0.6915 |
| 1 | 0.8413 |
| 1.5 | 0.9332 |
| 2 | 0.9772 |
| 2.5 | 0.9938 |
| 3 | 0.9986 |

sensitive to moisture. So with the slightest change in the monthly rainfall, immediately above zero, and if negative, to go below zero. SPI, long-term, reflecting the drought better and SPI values for this time scale, with floods, water levels in surface and ground water related.

Naturally, the SPI represents a *Z*-score or numerical standard deviation above or below the average for an event, although for a short-time scales as long as the distribution of precipitation, skewness is not quite correct.

According to, is worthy of mean and standard deviation indicator, is an appropriate way possible to compare different regions and stations will be significant.

Ekin and colleagues said they would choose and determine the spatial dimensions of time and drought cause problems, because if we have a drought significant estimates, not only must anomalies occurring in normal time scale for the future, but must be normalized with respect to the location. SPI index performs both cases. Index SPI, the location of the station is normally brought in when the frequency of the rainfall distribution combined with changes in estimates stations.

In addition, SPI time scale also comes as normal as it could be calculated at any time scale depending on the analyst's point of view. Thus, since the location is not the time scale of SPI limitation, in relation to the base period of the cumulative probability distribution parameters were estimated for its estimates (Tables 1 and 2).

### 2.2.2 The Use of Satellite Images to Extract Useful Data

In order to identify agricultural drought in the study area Landsat ETM+ sensor image processing with a resolution of 30 m and the data obtained from it is used. Selection criteria each Landsat satellite image of the area, during which time imaging in a specified period (for experimental) should be a maximum of cultivated land.

This period, according to the survey Irrigation Network North East (Vays, Mollasani and Salamat) the operational time surveying agricultural land, for the cultivation of summer and autumn months to 15 September was August 15. Moreover, for cultivation in winter and spring (like wheat), 15 March to 15 April is that the images were taken within the period that the steps necessary correction and processing on the images below have been done:

1. Obtaining images Cincinnati 165,038 (including the study area)
2. Scene 165,038 corrects errors on original image using auxiliary images (due to error or error SLC-OFF gap bar on image)
3. Scene image slices 165,038 to determine the target range
4. Study of image classification algorithm to extract the desired class
5. Select training samples (Training Area) to an assortment of images
6. Different classification algorithms on the image and initial evaluation (Classification)
7. Final processing of classification results (Post Classification)
8. The conversion of raster to vector format data categorization automatically formats
9. Basic editing results in vector format (removal of internal voids)
10. Visual editing using a combination of auxiliary data separately for each layer to
11. The composition of the final results of editing each data layer in GIS

Normalized Difference Vegetation Index (NDVI)

The normalized difference vegetation index that is among the simplest vegetation indices used, in terms of both red and near-infrared band in the form of Eq. (2) is defined.

$$\mathrm{NDVI} = \frac{\mathrm{NIR} - R}{\mathrm{NIR} + R} \tag{2}$$

As its name suggests this index in the range of normal values between −1 and +1, which makes it easy to evaluate and monitor values. NDVI represent different levels of coverage are different. For example, NDVI values are between 0.05 and 0.1 to areas of sparse vegetation, the values between 0.1 and 0.5 to areas of common vegetation, 0.5 to areas of high vegetation.

In this study, the image of vegetation index (NDVI) data for the selection of training (Training Data) carefully to extract information from the image, depending on the number of classes intended use. The use of vegetation index image possible occurrence during election training data in the image to a minimum.

Because in many cases, if only to the visual interpretation of your original image content may be two different effects on color band combination used wrongly and

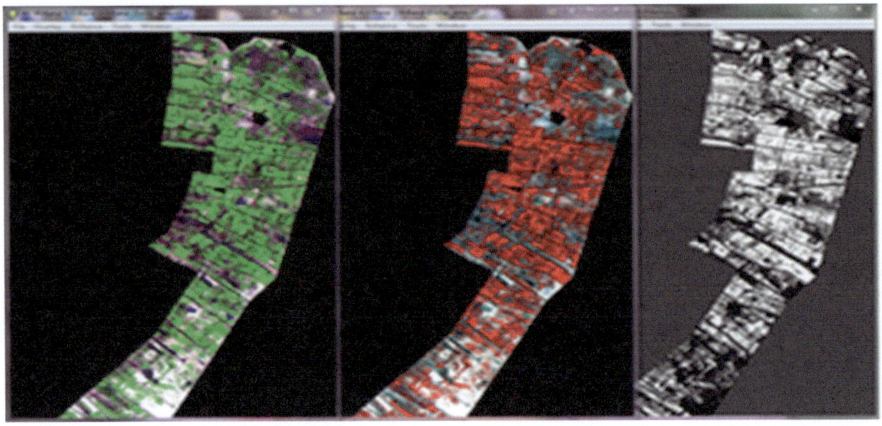

**Fig. 2** Comparison vegetation index image with the original image and a false image of the study area Digital methods of information extraction

mistakenly seen to be quite similar in the same class are described algorithm that is an important step in the implementation of the algorithm and its evaluation, the results will be wrong. Therefore, the use of vegetation index greater confidence of visual detection of different effects on the image to the user (Fig. 2).

Numerical methods for extracting information from satellite images are varied, and researchers are trying to provide optimal and efficient methods to extract information from satellite images.

Main methods of extracting information from satellite images (for digital) can be divided into four main categories:

1. Use of a threshold or cut investment
2. Visual computing
3. Classification
4. Segmentation

Since the classification method used in this research to the evaluation of these methods will be discussed and described other methods can be avoided.

Method of image classification (Classification)

Classification can be a decision-making process in which image data transmitted space specified class (Fatemi 2001).

In fact, the classification of a multispectral mapping from space to space complications. Space of a few spectral bands and each pixel in the image is created as a vector space is defined, each member of the vector of pixel values in a specific band. In the space of a few spectral (Multispectral Space) bands picture in front of each other, a multidimensional space (like an orthogonal coordinate system) to the user.

The most widely used classification methods of extracting information from satellite images.

There are a variety of classification methods, allows users to produce different types of information (such as use and land cover maps, calculate the volume density of vegetation, change detection, etc.) will be produced. Flexibility and high functionality compared to other classification methods have led to the most important methods used to extract information.

Some of the most important capabilities of non-spectral classification methods take advantage of the information and produce results more accurate and more complete, the production and the optimum use of information is one of the most important topics of interest to researchers.

Classification methods are known as regulated two categories (Supervised Classification) and unsupervised (Unsupervised Classification) are divided.

Supervised methods to basic information, such as the number of classes, properties, as well as some samples of each class, need it. In contrast to most methods for automatic unsupervised methods, it is not required to sample the pixel values, who do decide on their classification.

The cost and time needed information such low unsupervised methods and the accuracy of their results compared to the regulated low. Also, the results should be interpreted examined significant class.

In this study, the analysis tools available in the software ENVI classification of land or (TERCAT Analysis) were used for classification of adverse effects.

TERCAT tool outputs a class of pixels with similar spectral characteristics is. This class is defined either by the user or automatically by the classification algorithm is created. TERCAT all tools on the image classification algorithms implemented in the final. Finally, a classification or winning (Winner) between all methods of classification as the final output of the system. TERCAT, in fact, is the best class for each pixel based on the results of such minimum Euclidean distance algorithm (Minimum Distance), the minimum distance Mahalanobis (Mahalanobis Distance), maximum likelihood (Maximum Likelihood), corresponding spectral (Match Filtering).

### 2.2.3 The Use of GIS for Spatial Data Analysis

After processing the satellite images of the area using remote sensing software (ENVI), the data transfer to the Arc GIS software for computational analysis.

In this software to perform some computational analyses on the layers of information pertaining to each year using ArcMap, ENVI's classification layer transferred from this application, using save tool in ArcMap, save this layer. New layers created and stored, is a layer that can be applied in GIS software, and the operation is found. On this layer was operated, such as Reclassify, Raster to Polygon, Intersect, Eliminate, Dissolve, Multipart to Single part, in the Arc Map.

**Table 3** Status of wheat in the study area based on satellite image processing and analysis by GIS

| The area of cultivated land and uncultivated (in ha) (based on satellite image processing) | | | | | Status lands In terms of cultivate |
|---|---|---|---|---|---|
| Crop year | | | | | |
| 2008–2009 | 2007–2008 | 2006–2007 | 2005–2006 | 2004–2005 | |
| 3299 | 928 | 2890 | 3145 | 3303 | Wheat |
| 490 | 819 | 157 | 450 | 218 | Other agricultural products |
| 8035 | 10,078 | 8778 | 8229 | 8303 | Uncultivated land |

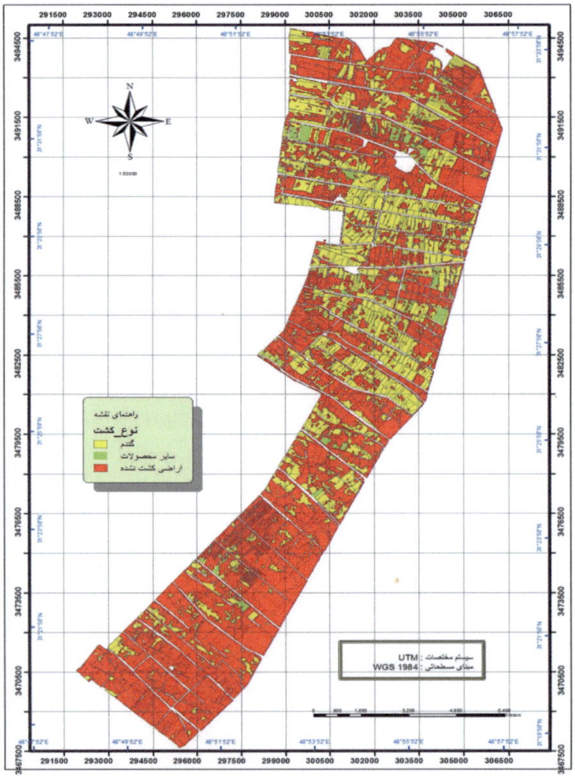


**Fig. 3** Cultivation of land in the area of study 2004–2005

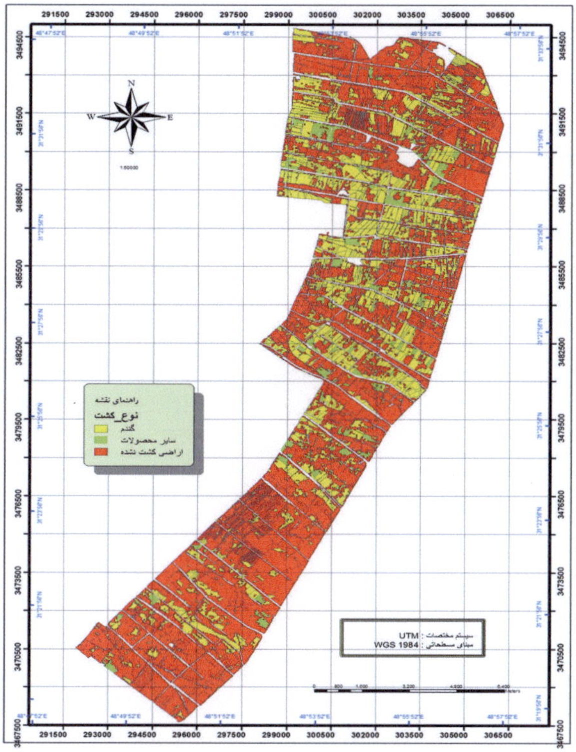


**Fig. 4** Cultivation of land in the area of study 2005–2006

The cadastral map layer of pure area network (arable land parts) that the DWG format to SHP (the format GIS) has become the ArcMap environment was recalled.

By adopting a land cadastre map and the map classification according to the database can be associated with the land situation in their cadastral map that has been cultivated wheat classification can be seen on the map and outputs calculated based on the map classification from satellite images can be extracted.

In ArcMap using appropriate means, according to the research, the analysis of spatial data (land parcels) were performed which leads to the final output of the project was analyzed as in Table 3 the results for all products in the region and the wheat crop is shown.

Figures 3, 4, 5, 6 and 7 in the process of shifting cultivation North East Irrigation in the period of 2004–2009 (four winter-spring periods) the satellite image processing and computational analysis obtained by GIS for wheat and other crops in the irrigation network is shown.

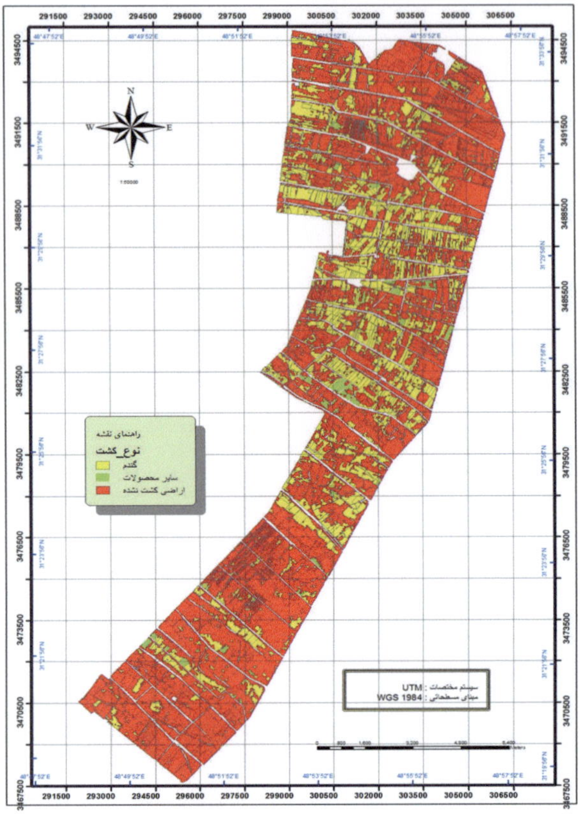


**Fig. 5** Cultivation of land in the area of study 2006–2007

Estimates of crop yield loss compared to the same period of cultivation in previous years, due to the occurrence of drought in the region, with the help of satellite imagery and use GIS takes place, authorities to understand that to provide appropriate management practices for water management in agriculture especially in irrigation and drainage networks cultivation of Kind of Blue was mainly to show greater sensitivity.

In the analysis and management of climate-related issues in the case before and during crises, such as drought, new technologies, such as remote sensing (RS) and geographic information system (GIS) should be used.

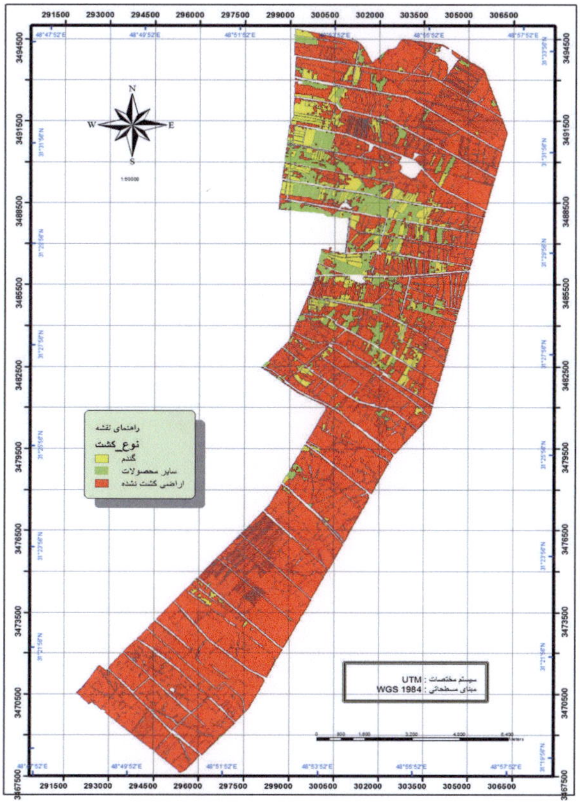


**Fig. 6** Cultivation of land in the area of study 2007–2008

# 3 Conclusion

The obtained results were then compared with the analyzed data obtained from GIS imagery and its juxtaposition using SPI indices showed a correlation between the occurrence of drought in the study period and a decrease of irrigated wheat yield. With due consideration to the repetition and duration of the drought in the future, it is necessary that drought preventive management strategies be applied in the agricultural sector and in addition to the optimal use of water in the irrigation network, modern irrigation techniques be applied in the farms. What more, it is proposed that existing agricultural patterns be changed and that a sturdier and more resistant variety be used during low water periods.

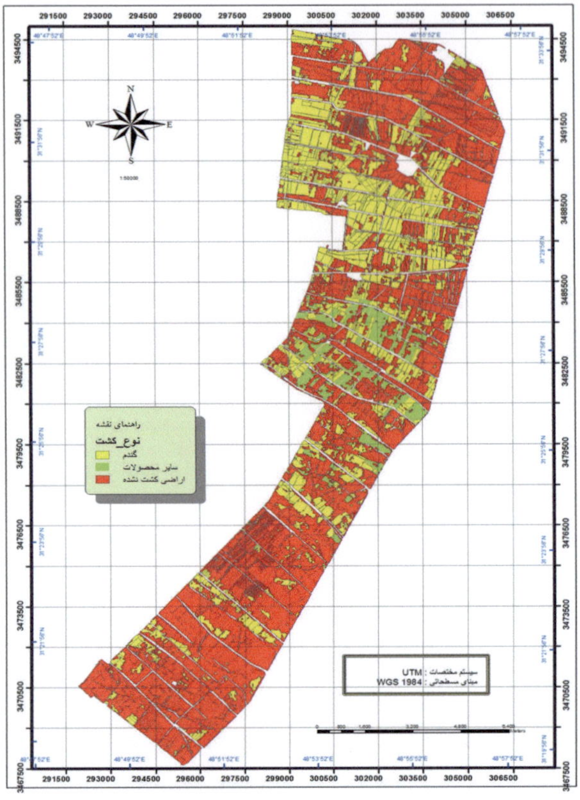


**Fig. 7** Cultivation of land in the area of study 2008–2009

# References

Hejazizadeh (2010) An introduction about Drought and The Indexes of Drought, The first Edition, Smart Publications (In Persian)

Fatemi S (2001) The principles of Remote Sensing, The Second Edition, Azadeh Publications (In Persian)

# Part III
# Smart City

# Communicating Disaster Risk Reduction Through Web-Map Applications

**Ali Asgary and Daryoush Kari**

**Abstract** Significant number of disaster risk reduction projects is implemented around the world. Each of these projects can provide guidelines for future projects and can be adopted by other communities if their information is properly communicated and shared. These projects can also become a useful source of knowledge for teaching and training of future disaster risk reduction professionals. This chapter describes a disaster risk reduction web mapping project called MapDRR that uses ESRI Webapp Builder to map more than 600 disaster risk reduction projects around the world. This application uses advances in web mapping to add more values to the existing online information about the disaster risk reduction projects. MapDRR covers a diverse range of DRR cases implemented in many different countries to reduce the risk of different types of disasters at local, regional, national, and international levels. MapDRR provides both textual and contextual (visual) information to users by zooming into each project. MapDRR is enhanced regularly to include more DRR cases, countries.

**Keywords** MapDRR · Disaster risk reduction · ArcGIS online · Webapp builder · Hazard mitigation

## 1 Introduction

Aiming to reduce disaster risks, governmental, non-governmental, private sector, and international agencies are undertaking large number of projects around the world. These projects not only reduce the risks to their intended beneficiaries, they

A. Asgary (✉)
Disaster and Emergency Management, School of Administrative Studies,
York University, Toronto, Canada
e-mail: asgary@yorku.ca

D. Kari
Communication & Culture Studies, York University, Toronto, Canada
e-mail: DKari@yorku.ca



S. Pirasteh and J. Li (eds.), *Global Changes and Natural Disaster Management: Geo-information Technologies*, DOI 10.1007/978-3-319-51844-2_7

generate a source of knowledge for other communities and countries if adequately communicated. Transfer of this knowledge to other communities and generations is very important for enhancement of DRR programs internationally (Bradley et al. 2011). Each of these projects can be viewed as a unique symbol of disaster risk reduction knowledge and action and often best practices and as such they can provide ideas and initiatives to other agencies in other communities who are trying to find and implement similar projects. While valuable information about some of these projects is available online, it is very scattered, textual, not easily accessible and identifiable.

Efforts have been made to create databases and online portals of Disaster Risk Reductions. One example of such efforts is the Disaster Risk Reduction Project Portal for Asia and the Pacific (www.drrprojects.net). This portal has been developed by the International Strategy for Disaster Risk Reduction in partnership with the Asia Partnership on Disaster Reduction (IAP) and funding support of the Asian Development Bank (ADB). This portal has been executed by the Asian Disaster Preparedness Center (ADPC). Disaster Risk Reductions portal provides information on all DRR projects in Asia and the Pacific region to increase collaboration and cooperation on conceptualization, planning, and programming on DRR among different stakeholders. The DRR Project Portal provides users with customizable tables, graphs, and maps of DRR projects. Data is provided by registered organizations that have implemented the DRR projects. The DRR Project Portal was developed using the Sahana Eden Open Source Disaster Management Platform (http://sahanafoundation.org/). This portal has been used in disaster risk reduction analysis (ISDR 2011). Built with ESRI Webmap App Builder, MapDRR provides more mapping and map-based search and analytical capabilities and as such differs very much from the disaster risk reductions portal. MapDRR can provide valuable information to DRR experts, practitioners, planners, researchers, and students.

Most DRR projects are location based. They are either carried out in a particular location (point) such a dam or bridge or in an area such as a neighborhood, city, or county. This provides an opportunity to map the projects. In recent years, a number of technologies have been introduced that provide easy-to-develop and easy-to-use interactive and participatory mapping systems that can be used to communicate DRR project more broadly. These tools enable us to design and create DRR project databases to better introduce these projects to interested individuals and organizations.

While portals like Disaster Risk Reduction Project Portal for Asia and the Pacific provide a good database with search, analysis, and mapping tools for the Asia and Pacific Region, the main goal of the MapDRR is to develop an alternative approach in mapping, visualizing, and reporting DRR projects using the power of ESRI ArcGIS Online Webapp Builder . As such we cover those DRR projects that have spatial attributes (i.e., latitude and longitude). The key advantages of this approach are: (1) Based on ESRI Base Maps, MapDRR provides more mapping contexts where users can zoom in and change base maps (satellite, terrain, national geographic, etc.) and see where exactly the DRR projects are being implemented relative to the hazards; (2) Using ESRI ArcGIS Online, spatial analysis tools can be

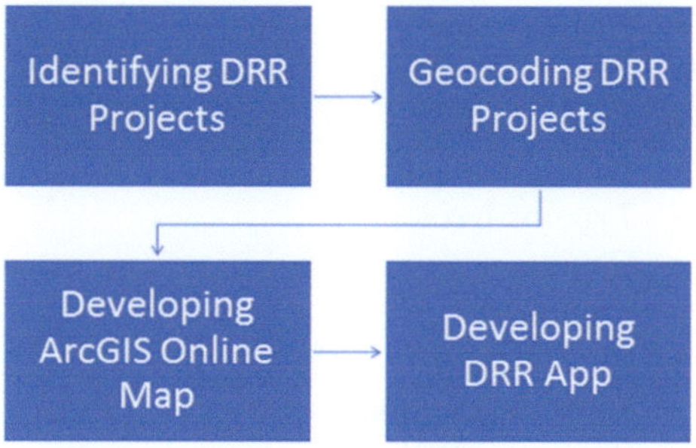


**Fig. 1** MapDRR development steps

added to the maps that will allow users to apply some basic mapping analysis; (3) Being in the cloud, there is no cost for these services to the users; and (4) the database can be easily shared and integrated in other applications.

This chapter describes the first phase of this educational and research initiative that has been started at York University to map world's DRR projects. Mapping these projects and making them more accessible can be a useful source for DRR professionals who are trying to find such measures and apply them in their own contexts. It also enables students to learn about different types of hazard and risk mitigation measures and their management processes.

The following steps have been taken in designing and developing the MapDRR:

1. Identifying disaster risk reduction projects
2. Geocoding DRR Projects
3. Mapping DRR projects Map and ArcGIS Online Webapp Builder (Fig. 1).

The rest of this chapter is organized as follows: Sect. 2 explains the MapDRR data development. Section 3 describes the contents of the current dataset. Section 4 briefly introduces the MapDRR application with some discussions. Section 5 concludes the chapter with some recommendations.

## 2 MapDRR Design and Development

Online resources were used as the main source to identify potential DRR projects. We applied a large number of basic keywords (i.e., "disaster risk reduction project", "hazard mitigation project", "risk mitigation project", etc.) and some very specific keywords (i.e., "flood disaster reduction project", "hurricane risk reduction", "earthquake risk reduction", etc.) to find the DRR projects. Considering our limited human resources and time, we identified about 630 projects that met our criteria. Our primary objective was to have a diverse sample DRR projects in terms of geographic distribution, type of DRR project, and type of hazard. We then identified a list of key attributes for the projects that need to be added as fields in the database. The following attributes were finally selected, considering the availability of information and their usefulness for the end users:

- Project name
- Project location
- Type of project
- Hazard category
- Type of hazard
- Country
- Continent
- City/Locality
- Short description
- Reference/Website
- Image

"Project name" represents the official that has been used for the project or activity. In most cases each project or activity has its unique name. Examples include: Dust storm mitigation project; Gretna flood control; Padiyapelella Town landslide mitigation; Mexico's battle against desertification; and Anti-Hail Cannon. Type of project refers to the type of mitigation or risk reduction measures. As such mitigation or risk reduction projects have been classified in terms of structural and nonstructural projects. This field allows users of the application to classify and query the DRR projects and view different projects in their own geographic contexts. For most structural projects the ability to find and zoom to them enables users to see the projects on the satellite images. Hazard category identifies DRR projects based on the hazards that they are trying to mitigate or reduce. We have used major hazard categories including natural; technological, and intentional/civic. Type of hazard(s) shows the type of hazard that the project has targeted. In most cases, DRR projects target single hazards, but when multiple hazards have been mentioned, the word "multi-hazard" has been used. Locational fields such as city/province, country, and continent have also been added to identify the overall location of the projects which will be very useful for creating charts, queries, and thematic maps.

Since we were interested in mapping the exact location of the DRR projects, we needed to find either the latitude and longitude or the full address of each project. It was a challenging task for some cases. A "Location" field was added to the database that contains the latitude and longitude of the projects separated by a comma in one field (for example: '29.295916, −94.804541'). While this information is critical for mapping DRR projects, it also helps in making connections between different projects and their local contexts. For example, some of the structural risk mitigation projects are visible in satellite imageries (or street views in Google map). In these cases, users can view these projects on their implementation site.

"Short Description" provides a summary for each project based on the available information in online sources. The "Reference" field takes the users to the original source for the project information if available. Finally, image field stores and displays an image of the projects upon availability.

### *2.1 Mapping DRR Projects with ArcGIS Online Webapp Builder*

Building on an earlier Google map version, this project (Asgary and Kari 2015), we developed an ArcGIS Webapp Builder version of the MapDRR project.

## 3 Disaster Risk Reduction Projects in the MapDRR Database

Currently, the main database includes about 650 DRR projects. They are being added to the MapDRR after careful review. Although more than half of the records are from the North America (55.1%), there are considerable numbers of projects from other continents in the database and we are aiming to increase this number. Figure 2 shows the current distribution of DRR projects in the MapDRR database.

The majority of the MapDRR project database records are related to the natural hazard category (87%). Only 11% of the records are technological, and 2% are intentional/civil disaster risk reduction projects (Fig. 3).

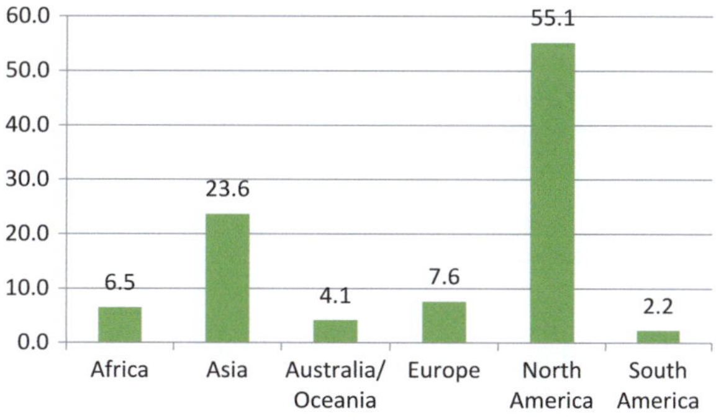


**Fig. 2** Distribution of DRR projects in the MapDRR database

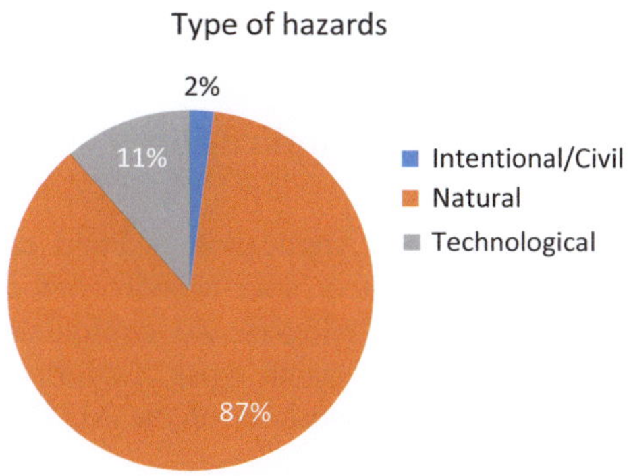


**Fig. 3** Distribution of DRR projects in term of hazard type

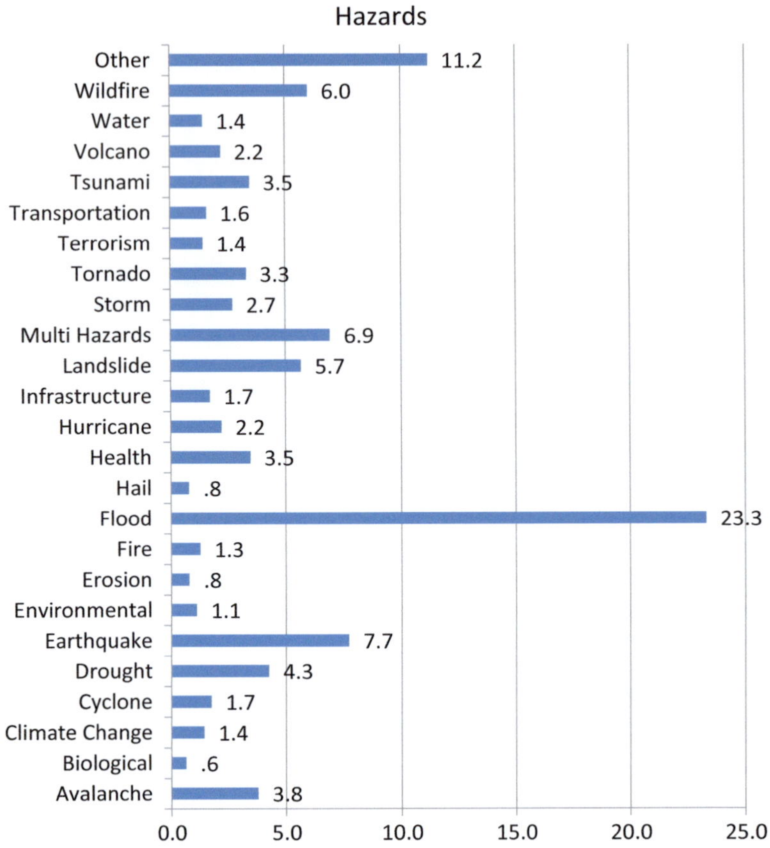


**Fig. 4** Distribution of DRR projects in term of hazard type

Figure 4 provides some detailed information about the DRR projects in terms of the type of hazards that they are dealing with. The current version of database covers a large number of different hazards. However, as expected significant number of the projects are flood-related projects (23%).

## 4 MapDRR Application

The "MapDRR" Application is build using the ERSI "Webmap App Builder" which is an intuitive, user-friendly, and device-independent platform. The use of readily available widgets and configurable themes provides the opportunity to create and extract various types of information from the data layers. Figures 5 and 6 provide sample snapshots of the MAP DRR project in ESRI Webmap Builder Application.

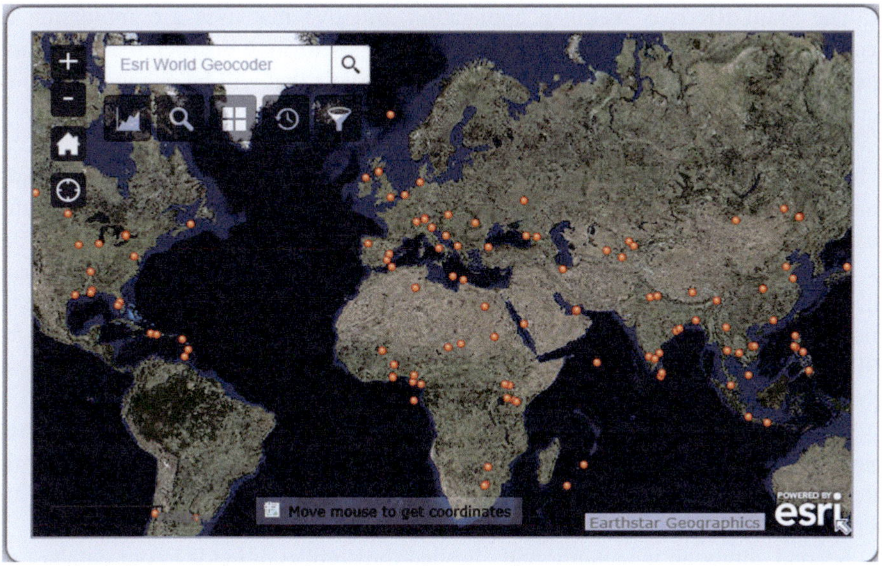


**Fig. 5** DRR projects in ArcGIS online Web app Builder

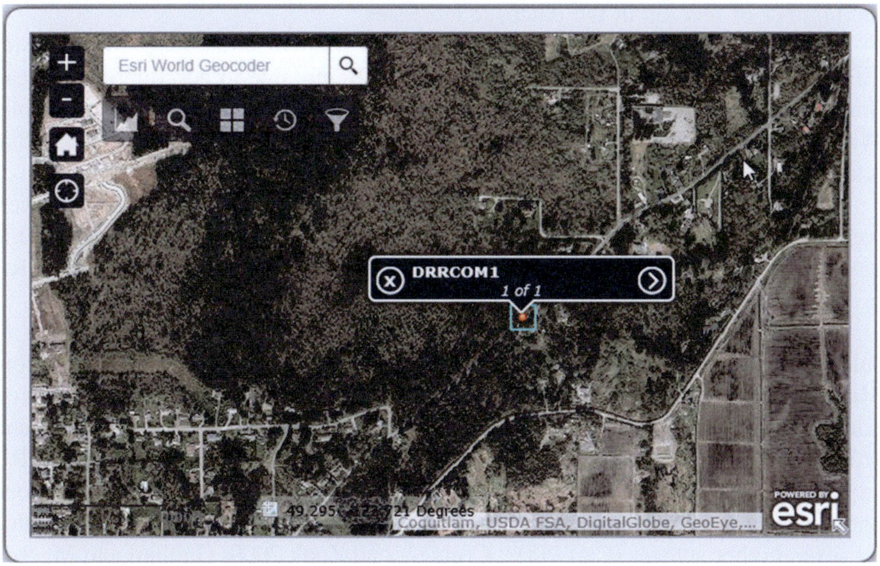


**Fig. 6** Sample disaster risk reduction project (Forest fire risk reduction project in British Columbia, Canada)

The MapDRR is a Webmap application initiative that uses a combination of recent mapping and cloud computing technologies to map disaster risk reduction projects around the world. This is an ongoing project that will be enhanced both in terms of its data and mapping capabilities. MapDRR can be very helpful for a variety of users including students, researchers, practitioners, and policy-makers who are interested in learning about different types of DRR projects that are being undertaken globally. It not only maps the DRR projects, but also provides useful textual and visual information to the users by bringing together the information that is publicly available in an easy-to-understand-and-navigate approach. Users can also perform analysis on the data through built-in analytical tools. It should be mentioned that developing a comprehensive and complete mapping system for DRR projects is not an easy task and faces many challenges. Some of these challenges and key areas of improvements are discussed here.

First, many DRR projects, especially those that have been implemented in the past or are being implemented by local governments in different parts of the world are not reported online (the main source for the MapDRR). Even if they are reported, they are not reported in English, the current language of MapDRR.

Second, although we intentionally limited the number of fields in our database to make it easier for the information providers and the end users, it is possible to add more fields to the database as required. However, adding more fields requires more database management system capacities. This might become a challenging task as the number of records in the database increases.

Third, the "Project Description" field in the MapDRR database contains significant textual information about each project. Being able to add queries that can filter projects based on the information provided in this field could be very valuable. We aim to add this functionality to the MapDRR so that users can find projects that have certain types of information in their descriptions.

Fourth, to receive the maximum benefits from creating the MapDRR base layer, we will add more features to MapDRR application. These features will allow the users to generate and view thematic maps, graphs, and charts for their selected DRR projects.

Fifth, to make MapDRR a participatory mapping tool, we will allow users to submit their DRR projects to be added to the database after verification. In addition, MapDRR can add a section on the database that collects and publishes users' comments about the viewed DRR projects. We are hoping that the participatory and crowdsourcing attributes of the MapDRR encourage more practitioners and DRR experts to provide DRR records and input to the database as highlighted by Doge and Kitchin (2013). This, of course, has its own challenges as discussed and argued by several other authors (Carr 2007; Crutcher and Zook 2009; Gaillard and Maceda 2009; Cadag and Gaillard 2012; de Vries et al. 2013).

Finally, we are developing specific symbols for each DRR project types to be replaced with the default ArcGIS online symbols. This will help users to visually identify the DRR projects on the App window much faster and easier.

## 5 Conclusion

MapDRR adds many additional benefits to the existing information that are available about DRR projects on the cloud. It provides DRR experts and researchers additional tool to easily find, analyze, and map DRRPs that have been implemented or are being implemented around the world. It will allow the users to compare different projects in different locations and learn from them using this interactive database and mapping tool.

There are many obvious improvements that need to be made to MapDRR, starting from adding more DRR records, structuring the database and adding more fields to the MapDRR base layer, adding more spatial analysis functionalities, query options, and users input to the MapDRR. However, our plan is to provide useful, reliable, and added value information and prioritize future extensions of MapDRR.

**Acknowledgments** Authors would like to acknowledge the following York University Students for their efforts in collecting the initial disaster risk reduction projects: Amina Mohammad Ali, Arham Arshad, Adam Witts, Ahmad Chaudhry, Andrea Nguyen, Colette Clarke, Dan Gutin, Toroptsev Denys, Fatemeh Nourinejad, Gita Javaheri, Hema Bahl, Hilal Faizal, Jean Kadenhe, Kevin Mak, Mohand Radi, Muhammad Mirza, Nima Sarjahani, Oliver Sandford, Paolo Di Rezze, Rudo Kadenhe, Rumaila Soin, Ryan Lucknauth, Sadegh Taheri, Sara Akib, Shahzaib Khalid, Shivam Kapoor, Toan Ly Ngo, Zaigham Imran, Tyler Willson, Uzma Bhatti, Vilma Lopez Luna, AllenRenee, and Maryanne Bendell.

## References

Asgary A, Kari D (2015) Mapping disaster risk reduction (MapDRR). The state of DRR at the local level a 2015 report on the patterns of disaster risk reduction actions at local level. UNISDR

Bradley ES, Roberts DA, Dennison PE, Green RO, Eastwood M, Lundeen SR, McCubbin IB, Leifer I (2011) Google Earth and Google Fusion tables in support of time-critical collaboration: mapping the deepwater horizon oil spill with the AVIRIS airborne spectrometer. Earth Sci Inform 4:169–179

Cadag JRD, Gaillard JC (2012) Integrating knowledge and actions in disaster risk reduction: the contribution of participatory mapping. Area 44:100–109

Carr NG (2007) The ignorance of crowds. Strategy Bus Mag, *47*, 1–5

Crutcher M, Zook M (2009) Placemarks and waterlines: racialized cyberscapes in post-Katrina Google Earth. Geoforum 40:523–534

de Vries S, Buijs AE, Langers F, Farjon H, van Hinsberg A, Sijtsma FJ (2013) Measuring the attractiveness of Dutch landscapes: identifying national hotspots of highly valued places using Google Maps. Applied Geography 45:220–229

Dodge M, Kitchin R (2013) Crowdsourced cartography: mapping experience and knowledge. Environ Planning A 45:19–36

Gaillard JC, Maceda EA (2009) Participatory 3-dimensional mapping for disaster risk reduction. Participatory Learn Action 60:109–118

ISDR (2011) At the cross roads: climate change adaptation and disaster risk reduction in Asia and Pacific. UNISDR Asia and Pacific Secretariat

# Mapping Sand Dune Fields in Abu Dhabi Emirate Over the Period of 1992–2013 Using Landsat Data

**N. Saleous, S. Issa and R. Saeed**

**Abstract** Up to 90% of the United Arab Emirates' (UAE) surface is covered by sand dunes and intervening inter-dune belts. The country is severely affected by problems related to sand dunes movement and Aeolian deposits, recognized as a major contributor to desertification. This study discusses the use of publicly available Landsat TM and ETM+ data to detect sand dunes fields and enable monitoring of their movements in the Emirate of Abu Dhabi, UAE. The study focuses on developing a classification approach and applying it to historical Landsat data to produce consistent Land cover maps useable in subsequent change detection studies. Landsat scenes acquired over the period 1992–2013 are used to evaluate different multispectral classification approaches and determine the accuracy of resulting maps. The methodology uses several configurations of supervised classification techniques that include different band combinations to determine those that produce the highest accuracy in mapping the predominant land cover classes in the area. Results indicate that the tested configurations exhibit an unacceptable level of confusion in detecting the built-up class and that the use of surface reflectance as input to supervised classification yields adequate results for sand detection. All configurations also exhibit a certain level of confusion between sparse vegetation and other classes. The use of a vegetation index as a discriminator helps improve the classification accuracy.

**Keywords** Sand dunes · Classification approaches · Landsat · Monitoring · Vegetation index · Change analysis · UAE

N. Saleous (✉)
Geography Department, United Arab Emirates University,
17551, Al-Ain, United Arab Emirates
e-mail: nazmi.saleous@uaeu.ac.ae

S. Issa
Geology Department, United Arab Emirates University,
17551, Al-Ain, United Arab Emirates

R. Saeed
RS/GIS Postgrad, United Arab Emirates University,
17551, Al-Ain, United Arab Emirates

 
S. Pirasteh and J. Li (eds.), *Global Changes and Natural Disaster Management: Geo-information Technologies*, DOI 10.1007/978-3-319-51844-2_8

# 1 Introduction

A major part of the United Arab Emirates is covered by sand dunes and sand sheets. With the rapid growth that the country witnessed after the discovery of oil in 1968, major urbanization and farming projects were undertaken throughout the country some in the heart of the sand sea. Sand dunes and their movements present a serious sand encroachment threat to these structures. This study focuses on the mapping of sand dune fields and assessing their changes over the period 1992–2013 in the largest emirate of UAE using Landsat imagery. Availability of similar spatial and spectral characteristics of the Landsat TM and ETM+ sensors during the study period will ensure that they will provide consistent and reliable imagery that can be used in this study (Masek et al. 2001; Teillet et al. 2001).

Many previous studies explored the use of satellite imagery for detecting sand dunes and sand sheets. Different approaches to extract sand features were studied and evaluated. Some authors used enhancement and filtering techniques followed by digitizing sand fields (Janke 2002; Al-Dabi et al. 1997). Others used the spectral characteristics of sand to detect it through a multispectral classifications process (Hutchison 1982; Tuller 1987; Smith et al. 1990; Janke 2002).

Some of the approaches used in land cover classification suggest supplementing the spectral information with spatial information through the use of texture (Chang and Park 2006). Others suggest a transformation of input data through the use of Principal Component Analysis (PCA) to increase the classification accuracy (Rajesh 2008; Fung and LeDrew 1987).

In this study, we examine different approaches to create, detect, and map sand using Landsat imagery collected in three different years: 1992, 2002, and 2013. We apply the selected approach to create land cover and sand/non-sand maps over the study area for the three different dates. We assess their accuracy using higher resolution imagery and store them in vector format for use in change detection studies.

# 2 Study Area

The study area (Fig. 1) is located along the southern coast of the Arabian Gulf, between 22° 40″ and 25° North and 52° and 56° East. Approximately 85% of the Emirate is desert and can be divided into five different landforms: sand dunes (moving desert), inter-dune areas, coastal sabkhas, inland sabkhas, and exposed rock (Glennie 2001; Abu Dhabi Urban Planning Council 2007). Aeolian sand dunes are prominent in the UAE due to the country's climatological patterns, where the lack of rainfall and high summer temperatures, coupled with the scarcity of vegetation cover and strong wind patterns, all contribute to the suitable conditions that enhance sand dune movement (Glennie 2001; Al-Awadi 2004).

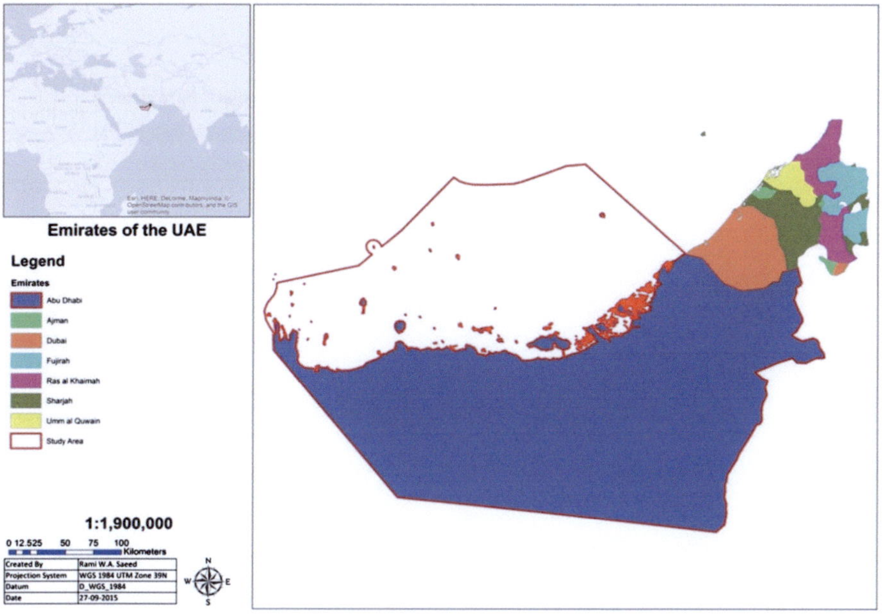


**Fig. 1** Map of the UAE with the study area delineated by the *red border*

The Emirate's climate is harsh with mainly two seasons. One is the hot summer (May–September) with rare rainfall where temperatures reach 48 °C; humidity ranges between 80 and 90% and sandstorms regularly strike between May and early July. The other is a mild winter where the temperature reaches 24 °C, where fog becomes more frequent, and little rainfall occurs more between November and February.

## 3 Methodology

### *3.1 Data*

Six Landsat scenes defined by paths 160–162 and rows 43–44 of the WRS-2 frame are required to cover the selected study area. A set of these scenes for each of the study dates 1992, 2002, and 2013, are acquired from the Landsat Ecosystem Disturbance Adaptive Processing System (LEDAPS) Climate Data Record (CDR) repository at the United States Geological Survey (USGS) (Masek et al. 2006). Their characteristics are summarized in Table 1. The radiometrically corrected surface reflectance product included in the collected data set are processed using Exelis Visual Information System ENVI 5.3 image processing software to extract and build a time series of the sand class in the study area.

**Table 1** Characteristics of acquired Landsat scenes

<table>
<tr><th rowspan="2">Row</th><th colspan="3">Path<br>Date</th><th rowspan="2">Resolution</th><th rowspan="2">Parameter</th><th rowspan="2">Sensor<br>Platform</th></tr>
<tr><th>162</th><th>161</th><th>160</th></tr>
<tr><td rowspan="3">43</td><td>04-06-1992</td><td>05-06-1992</td><td>14-06-1992</td><td rowspan="6">30 meters</td><td rowspan="6">CDR<br>Surface<br>Reflectance</td><td>TM<br>Landsat<br>4</td></tr>
<tr><td>03-07-2002</td><td>10-07-2002</td><td>12-07-2002</td><td>ETM+<br>Landsat<br>7</td></tr>
<tr><td>01-08-2013</td><td>25-07-2013</td><td>18-07-2013</td><td>OLI<br>Landsat<br>8</td></tr>
<tr><td rowspan="3">44</td><td>04-06-1992</td><td>05-06-1992</td><td>14-06-1992</td><td></td></tr>
<tr><td>03-07-2002</td><td>10-07-2002</td><td>12-07-2002</td><td></td></tr>
<tr><td>01-08-2013</td><td>25-07-2013</td><td>18-07-2013</td><td></td></tr>
<tr><td>Datum<br>Projection</td><td>WGS 1984<br>UTM Zone 39 N</td><td>WGS 1984<br>UTM Zone 39 N</td><td>WGS 1984<br>UTM Zone 40 N</td><td></td><td></td><td></td></tr>
</table>

**Table 2** Characteristics of secondary data used in the study

| Data type | Date | Spatial resolution (m) | Purpose |
|---|---|---|---|
| Rapid eye | 2013 | 5 | Accuracy assessment |
| IKONOS | 2003 | 1 | Accuracy assessment |
| SPOT panchromatic | 1986 | 10 | Accuracy assessment |
| Abu Dhabi Emirate boundary shapefile | 2013 | – | Delimit study area |

Additionally, a set of secondary data is needed in the study to delimit the study area and to perform accuracy assessment of the created land cover maps. They are listed in Table 2.

## 3.2 *Methods*

The approach used in this study is summarized in Fig. 2. It includes mosaicking individual Landsat scenes for each study year and creating surface reflectance subsets of the study area that are subsequently used in the multispectral classification process. The approach also defines a classification scheme including

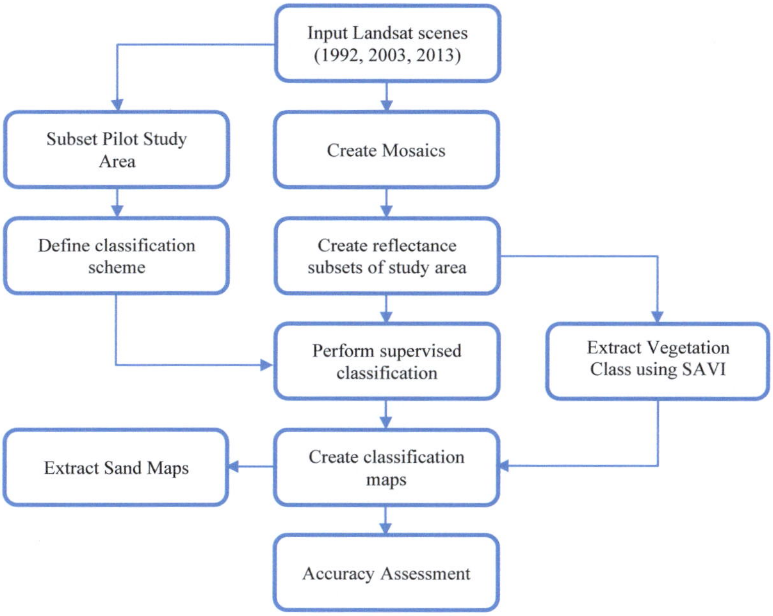


**Fig. 2** Methodology flowchart

important land cover classes in the study area that can be adequately derived from the available dataset. It calls for the selection of a smaller pilot study area where multiple configurations of inputs are used to identify detectable spectral clusters.

To improve the overall accuracy of the classification, we extract the vegetation class using a vegetation index approach prior to performing supervised classification using the surface reflectance mosaics of the study area. As a result, land cover maps are created for the three study years, and their accuracy is assessed using the confusion matrix technique. Finally, sand maps that can be used in analyzing sand changes are extracted (Fig. 2).

## *3.3 Extraction of Vegetation Class*

Since in arid regions, vegetation is usually sparse and can be spectrally confused with other cover types at Landsat spatial resolution, we opt to extract the vegetation class separately using a vegetation index approach. Given the sparse nature of vegetation with exposed soil-understory, we select the Soil Adjusted Vegetation Index (SAVI) to help identify vegetated areas. SAVI is calculated from the Red and Near-Infrared surface reflectance using the formula given in Eq. (1) (Huete 1988).

$$\mathrm{SAVI} = \frac{\mathrm{NIR} - \mathrm{RED}}{\mathrm{NIR} + \mathrm{RED} + L} * (1 + L), \tag{1}$$

where $L$ is the factor that highlights the difference between vegetation and soil and is set 0.5.

The analysis of the produced SAVI maps for the three study periods leads to determine the vegetation threshold to be 0.35. All pixels where SAVI is greater than 0.35 are labeled vegetation and are excluded from the subsequent classification process.

## 3.4 Classification Scheme

The classification scheme used in this study is based on a modified Level 1 Anderson scheme. Classes in the original scheme that do not occur in UAE are removed from the scheme. They are rangeland, forest land, tundra and perennial snow, and ice. As mentioned earlier, vegetation class (agricultural land) is extracted separately and is excluded from the classification process.

Due to the limited spatial resolution of Landsat data coupled with the heterogeneity of certain land cover classes such as the urban class, it cannot be detected accurately. To minimize detection errors, we test several configurations of input parameters to the clustering algorithms over a pilot study area to ensure that the selected classes can be identified using Landsat data. The pilot study area is limited to one Landsat scene that includes all desired classes. The test set of inputs are: (1) Texture entropy, (2) surface reflectance and texture, (3) surface reflectance and (4) three first components of the PCA. The results of the clustering using the ISODATA algorithm are presented in Fig. 3. Other than the texture entropy, the different input configuration yielded a substantial number of different clusters: 31 clusters in the case of surface reflectance, 41 clusters in the case of surface reflectance + texture and 295 clusters in the case of PCA. However, merging the clusters into information classes produces the same number of detectable classes for all three configurations such as in the example presented in Fig. 4. The classes that can be identified are: sand, vegetation (as derived from the vegetation index), water (deep water), intertidal zone (shallow water), wet soils and sabkhas, and exposed bedrock. The urban class is consistently confused with wet soils or exposed bedrock classes. Since the main focus of this study is sand, the errors in the detection of the urban class can be tolerated.

Since the different input configurations yield comparable detectable information classes, the configuration using the surface reflectance only is used in the classification process. A class separability analysis confirmed that the selected classes are spectrally identifiable.

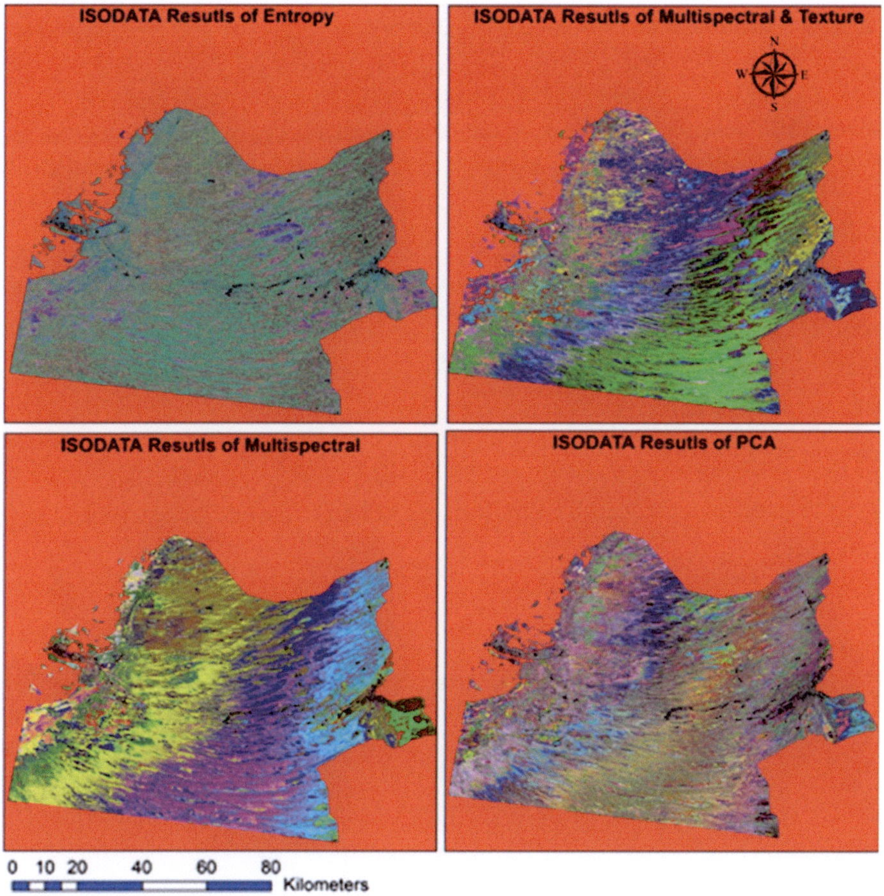


**Fig. 3** Results of the clustering algorithm using the four input configurations

## 3.5 *Classification and Accuracy Assessment*

Once the classification scheme defined and the vegetation class extracted and excluded through a masking process, a supervised classification approach using the surface reflectance as input is applied to map the remaining classes: sand, intertidal zone, wet soil/sabkha, and exposed bedrock. The supervised classification process uses 142 training sites that we selected randomly across the different classes. Speckles and random errors in the classified maps are then removed using clumping and sieving filter.

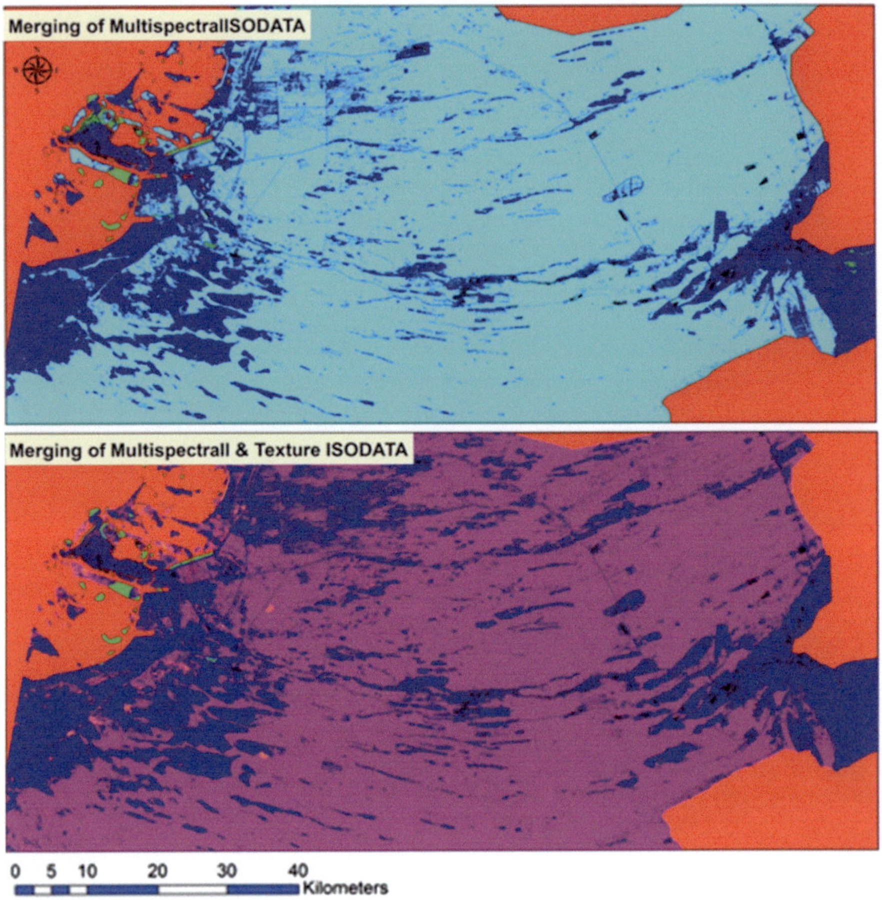


**Fig. 4** Results of merging the clusters for the surface reflectance and surface reflectance + texture cases

To assess the classification accuracy, confusion matrices for the three study years are created from a set of 94 assessment invariant sites. The sites are selected using a stratified random sampling to ensure all classes are covered. The invariant sites are selected using the available higher resolution imagery: 2013—5 m RapidEye; 2003—1 m IKONOS and 1986—10 m SPOT. Ensuring that the selected sites consistently represent the same land cover classes for these three dates gives us confidence in their invariability allowing us to use them in the assessment of three land cover maps we produce.

# 4 Results and Discussion

## *4.1 Land Cover and Sand/Non-sand Maps for Years 1992, 2002, and 2013*

The results of the classification process are presented in Fig. 5. Land cover maps for the years 1992, 2002, and 2013 show the distribution of the different land cover classes in the emirate. It is worth noting though that due to the limitations of the input data, the urban/built-up class is assigned to either wet soil/sabkha class or exposed bedrock class making the distribution of these classes inaccurate. However, since the main objective is to map sand dunes. The extracted classes are merged into two categories: sand and non-sand. The resulting sand/non-sand maps are also presented in Fig. 5. Table 3 summarizes the size of each one of the classes

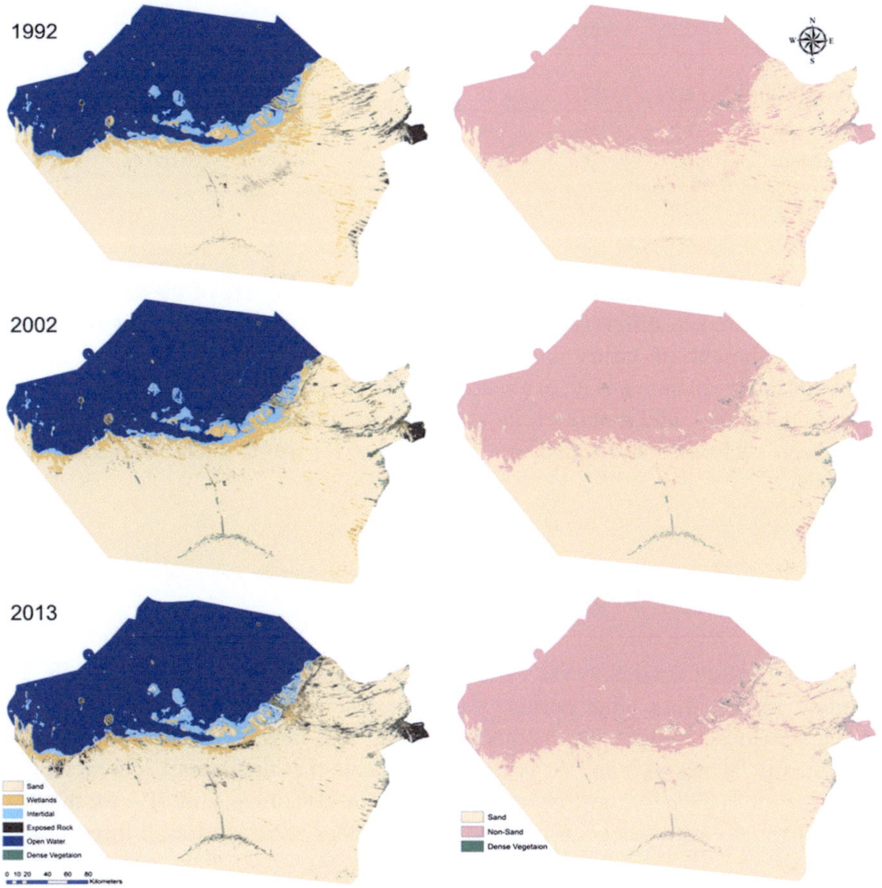


**Fig. 5** 1992, 2002, and 2013 land cover and sand maps

**Table 3** Size of the sand/non-sand classes for 1992, 2002, and 2013

| Class | 1992 ($km^2$) | 2002 ($km^2$) | 2013 ($km^2$) |
|---|---|---|---|
| Vegetation | 147 | 518 | 440 |
| Wet soil | 5168 | 4341 | 2651 |
| Intertidal | 3069 | 2788 | 3175 |
| Exposed bedrock | 1874 | 1864 | 3821 |
| Water | 32,172 | 32,873 | 32,678 |
| Non-sand (excluding water) | 10,258 | 9,511 | 10,087 |
| Sand | 52,186 | 52,732 | 52,848 |

**Table 4** Confusion matrix for 1992

| Class | Sand | Wet soil | Intertidal | Exposed bedrock | Open water | Vegetation | Total |
|---|---|---|---|---|---|---|---|
| Sand | 90.98 | 0 | 0 | 23.02 | 0 | 0 | 64.23 |
| Wet soil | 2.88 | 100 | 0 | 25.68 | 0 | 0 | 17.56 |
| Intertidal | 0 | 0 | 100 | 0 | 0 | 0 | 1.76 |
| Exposed bedrock | 6.14 | 0 | 0 | 51.30 | 0 | 0 | 10.56 |
| Open water | 0 | 0 | 0 | 0 | 100 | 0 | 4.59 |
| Vegetation | 0 | 0 | 0 | 0 | 0 | 100 | 1.29 |

for the three time periods. It highlights the changes in the size of the sand class that can be attributed to different factors including sand encroachment, urban growth, the establishment of farms and dredging. However, a more thorough change detection study, planned as an extension of the current study, is needed to assess the magnitude, extent, and nature of the change. Land cover maps created in this study have been vectorized and saved in a geodatabase for use in such change detection analysis.

## *4.2 Accuracy Assessment*

The classification accuracy was assessed using a confusion matrix approach. The matrices for the three time periods were constructed using the assessments sites selected as described in the methods section. The three confusion matrices are presented in Tables 4, 5 and 6. The results indicate that the overall accuracy of the classification is 87% for the 1992 map, 89% for the 2002 map, and 91% for the 2013 map. However, sand has been mapped with a higher accuracy for all three years.

**Table 5** Confusion matrix for 2002

| Class | Sand | Wet soil | Intertidal | Exposed bedrock | Open water | Vegetation | Total |
|---|---|---|---|---|---|---|---|
| Sand | 99.06 | 9.76 | 0 | 44.60 | 0 | 0 | 73.89 |
| Wet soil | 0.94 | 80.59 | 0 | 16.95 | 0 | 0 | 12.68 |
| Intertidal | 0 | 9.43 | 100 | 0 | 0 | 0 | 2.94 |
| Exposed bedrock | 0 | 0.18 | 0 | 38.44 | 0 | 0 | 4.52 |
| Open water | 0 | 0 | 0 | 0 | 100 | 0 | 4.65 |
| Vegetation | 0 | 0 | 0 | 0 | 0 | 100 | 1.31 |

**Table 6** Confusion matrix for 2013

| Class | Sand | Wet soil | Intertidal | Exposed bedrock | Open water | Vegetation | Total |
|---|---|---|---|---|---|---|---|
| Sand | 99.75 | 0 | 0 | 56.15 | 0 | 0 | 74.38 |
| Wet soil | 0.06 | 85.67 | 0.85 | 0.12 | 0 | 0 | 10.63 |
| Intertidal | 0 | 9.97 | 99.15 | 0 | 0 | 0 | 2.99 |
| Exposed bedrock | 0.19 | 4.37 | 0 | 43.73 | 0 | 0 | 6.10 |
| Open water | 0 | 0 | 0 | 0 | 100 | 0 | 4.60 |
| Vegetation | 0 | 0 | 0 | 0 | 0 | 100 | 1.30 |

# 5 Conclusion and Recommendations

Publicly available Landsat data were used successfully to map sand in the Emirate of Abu Dhabi for three periods spanning over two decades. The sand mapping approach used in this study relied on a robust supervised classification technique. The use of radiometrically corrected surface reflectance as input allowed the use of a consistent set of training sites across the three periods and paved the way to extend the study to other periods. Similarly, a set of invariant classification assessment sites was selected using high-resolution imagery. The levels of accuracy achieved in mapping the sand and non-sand classes make the produced maps useful for change detection and change trajectory studies. To that effect, the produced classification maps were vectorized and stored in a geodatabase to facilitate its use in a GIS environment. This research should be continued to derive, analyze sand change trajectories, and develop a model for its movement.

A major issue encountered in this study was related to the limitations of the moderate spatial resolution Landsat data in identifying accurately heterogeneous classes, such as urban areas. It is recommended to explore the use of higher resolution imagery to address this shortcoming.

## References

Abu Dhabi Urban Planning Council (2007) Abu Dhabi 2030: urban structure framework plan

Al-Awadi JMA (2004) Sand flow. Specialized book series, 1st edn. Kuwait Institute for Scientific Advancement, Directorate of Scientific Culture, State of Kuwait, p 216

Al-Dabi H, Koch M, Al-Sarawi M, El-Baz F (1997) Evolution of sand dune patterns in space and time in north-western Kuwait using Landsat images. J Arid Environ 36:15–24

Chang EM, Park K (2006) Feature extraction in an aerial photography of gimnyeong sand dune area by texture filtering. J Korean Geogr Soc 6:139–149

Fung T, LeDrew E (1987) Application of principal components analysis change detection. Photogram Eng Remote Sens 53:1649–1658

Glennie KW (2001) Evolution of the Emirates' land surface: an introduction. In: Al-Abed I, Hellyr P (eds) United Arab Emirates: a new perspective. Trident Press, London, pp 9–27

Huete AR (1988) A soil-adjusted vegetation index (SAVI). Remote Sens Environ 25:295–309

Hutchison CF (1982) Remote sensing of arid and semi-arid rangeland. International geoscience and remote sensing symposium, Munich, Federal Republic of Germany, 5.1–5.6

Janke JR (2002) An analysis of the current stability of the Dune Field at Great Sand Dunes National Monument using temporal TM imagery (1984–1998). Remote Sens Environ 83:488–497

Masek JG, Honzak M, Goward SN, Liu P, Pak E (2001) Landsat-7 ETM+ as an observatory for land cover: initial radiometric and geometric comparisons with Landsat-5 Thematic mapper. Remote Sens Environ 78:118–130

Masek JG, Vermote EF, Saleous N, Wolfe R, Hall FG, Huemmrich F, Gao F, Kutler J, Lim TK (2006) A Landsat surface reflectance data set for North America, 1990-2000. Geosci Remote Sens Lett 3:68–72

Rajesh HM (2008) Mapping proterozoic unconformity-related uranium deposits in the Rockhole area, Northern Territory, Australia using Landsat ETM+. Ore Geol Rev: 382–396

Smith MO, Ustin SL, Adams JB, Gillespie AR (1990) Vegetation in deserts: I. A regional measure of abundance from multispectral images. Remote Sens Environ 31:1–26

Teillet PM, Barker JL, Markham BL, Irish RR, Fedosejevs G, Storey JC (2001) Radiometric cross-calibration of the Landsat-7 ETM+ and Landsat-5 TM sensors based on tandem data sets. Remote Sens Environ 78:39–54

Tuller PT (1987) Remote Sensing science application in arid environment. Remote Sens Environ 23:143–154

# Spatiotemporal Analysis and Image Registration for Studying Growth of Transportation Infrastructure in Sharjah City, UAE

**Rami Al-Ruzouq, Khaled Hamad and Abdallah Shanableh**

**Abstract** Sharjah is the third largest and populous city in the United Arab Emirates (UAE). It is located along the northern coast of the Persian Gulf on the Arabian Peninsula. Throughout the past few decades, Sharjah City has witnessed massive growth in its urban area and infrastructure facilities. Transportation infrastructure, in particular, is a key indicator of growth and development of the city. Transportation is a vital component of the economy and prosperity of any city as it provides easy access to land, moves large volumes of people and vehicles, enables larger markets, and saves time and costs. Recent advances in satellite imagery, in terms of improved spatial and temporal resolutions, are allowing for efficient identification of change patterns and the prediction of areas of growth. This paper aims to quantify and analyze the spatial–temporal relationship between urban growth and the transportation development that took place at Sharjah City from 1976 until 2014. For growth detection and quantification, linear features extracted automatically from multi-temporal Landsat registered images were adopted as the basis of change detection where pixel-to-pixel subtraction has been implemented. Linear features were also chosen for image registration since they can be reliably extracted from imagery with significantly different geometric and radiometric properties. Digitized features of building and roads have been used as ground-truth of the adopted algorithm. Preliminary results show that the range of growth represented by linear features (building and roads) that occurred during the 1976–2014 period accounts for about 33% of the total area inside Sharjah City.

**Keywords** Transportation infrastructure · Analysis of growth patterns · Image registration

---

R. Al-Ruzouq (✉) · K. Hamad · A. Shanableh
Civil and Environmental Engineering Department,
University of Sharjah, Sharjah, United Arab Emirates
e-mail: ralruzouq@sharjah.ac.ae

 
S. Pirasteh and J. Li (eds.), *Global Changes and Natural Disaster Management: Geo-information Technologies*, DOI 10.1007/978-3-319-51844-2_9

# 1 Introduction

Transportation is key to promoting economic prosperity and urban development. Transportation system provides access to business activities, employment, education, and recreational opportunities. As such, the adequacy of transportation infrastructure, whether it is roadways or transit, has a great impact on any city's development and on maintaining socially acceptable levels of quality of life. As an urban area develops, its transportation system grows in parallel. In fact, the expansion of transportation infrastructure is often observed as synonymous to urban growth.

Over the past four decades, Sharjah City, the third largest city in the UAE, has witnessed a massive growth in its population, urban area, and infrastructure. For example, the population of Sharjah has increased almost tenfold between 1975 and 2005 (Wikipedia 2016). Since urban growth varies spatially, it is of interest to officials and researchers alike to quantify and trace the spatial growth of the City over time. The main purpose of this study is to detect urban infrastructure growth and development in Sharjah City by detecting and registering linear features in multi-temporal Landsat images.

Change detection is the process of identifying differences in the state of an object or phenomenon by observing it at different times (Singh 1989). Multi-spectral and multi-resolution images captured by various sensors were used to quantify changes at different times. Numerous change detection methods have been illustrated and described in the literature. These methods adopted different algorithms that basically depend on image subtraction, image ratio, change vector analysis, principle component analysis, neural network, or morphological mathematics (Al-Ruzouq and Habib 2012; Bruzzone and Prieto 2000; Dowman 1998; Singh 1989).

Image registration can be defined as the process that can be used in order to geometrically aligning multi-spectral and multi-resolution images where same features in different images are belonging to each other. This process required major elements. First; Registration primitives (points, lines or polygons), second; appropriate transformation function and matching strategy. Accurate image registration is essential prerequisite for accurate change detection where changes in multi-spectral and multi-resolution images belongs to real changes rather than changes resulted from miss registration process.

In this paper linear features extracted from roads, affine transformation function and Modified Iterated Hough Transform (MIHT) (Habib and Al-Ruzouq 2004; Habib and Kelley 2001a, b) were adopted as base for a registration methodology.

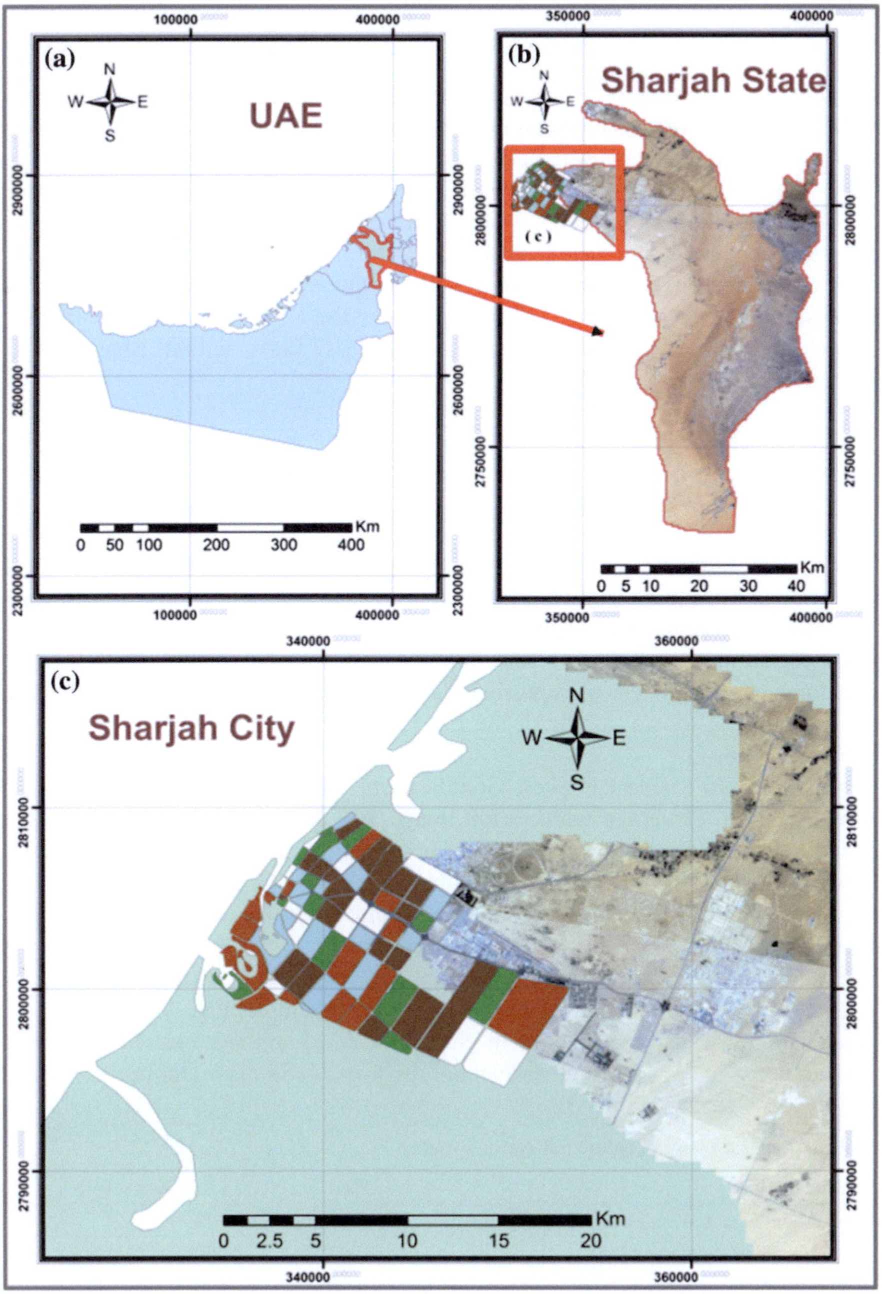


**Fig. 1** Location of the study area: UAE map, Landsat image of Emirate of Sharjah and the study area within Sharjah city

## 2 Study Area

Sharjah City is the largest city and the seat of the Emirate of Sharjah in the United Arab Emirates (UAE). It is located in a dry hot region, with daily high temperatures of 24–42 °C and daily main temperature of 18–34 °C. The rainy months are typically December to February, with an average rainfall of approximately 100 mm/year. The rainfall in dry years does not generally exceed 40 mm/year but can reach up to 200 mm/year in heavy-rain years. The City of Sharjah is located along the northern coast of the Persian Gulf on the Arabian Peninsula with central coordinates of 25.3°N 55.5°E. Figure 1a–c show the UAE map, Landsat image of Emirate of Sharjah, and the study area (136.27 $km^2$) within Sharjah City, respectively.

## 3 Geospatial Data

Multi-source, multi-resolution, and multi-temporal imagery for the City of Sharjah were used in this study. The diverse satellite images with different radiometric and geometric resolutions were captured in different years. These data will be used for image registration and change detection. Table 1 lists the source of each image including source and year of capturing the image as well as its spatial resolution in meters. Figure 2 shows sample of the Landsat images for the study area for years 1976 and 2010.

In addition to Landsat images, Google Earth images have been captured with high resolution. Coverage within each image was minimized during capturing to ensure high resolution images. Five images were captured and geo-referenced to cover the study area. Figure 3 shows sample of mosaic Google Earth images with 3.0 m ground resolution. This resolution would allow for accurate digitizing of linear features as described in the following section.

In addition to the above-described raster data, vector data, such as land use and roadway network data, have been prepared. As an example, Fig. 4 shows land-use classification for the study area in Sharjah City. Land-use classes considered include industrial, educational, residential, high-rise buildings, and mixed use.

**Table 1** Landsat images captured at various years

| Satellite | Year | Resolution (m) |
|---|---|---|
| Landsat multispectral scanner (MSS) | 1976 | 79 |
| Landsat thematic mapper (TM) | 1987 | 60 |
| Landsat thematic mapper (TM) | 2000 | 30 |
| Landsat thematic mapper (TM) | 2007 | 30 |
| Landsat enhanced thematic mapper (ETM) | 2010 | 15 |
| Google earth image | 2004 | 3 |
| Google earth image | 2014 | 3 |

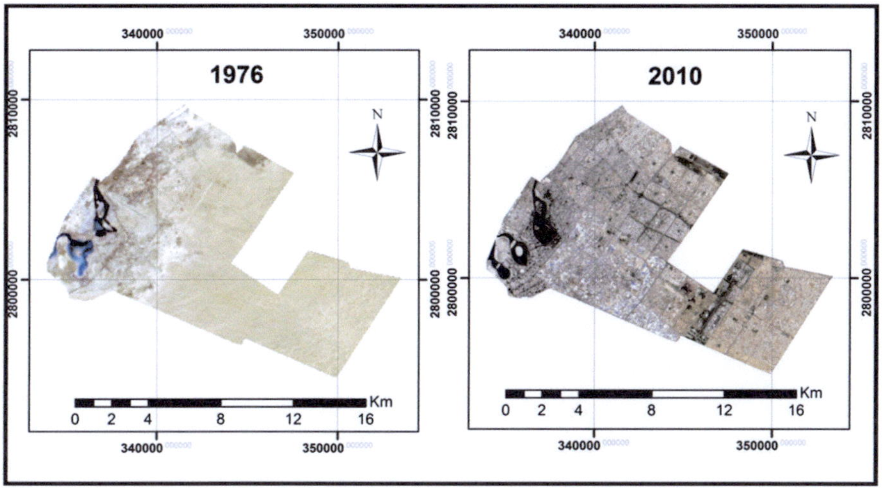


**Fig. 2** Landsat images of the study area for the years 1976 and 2010

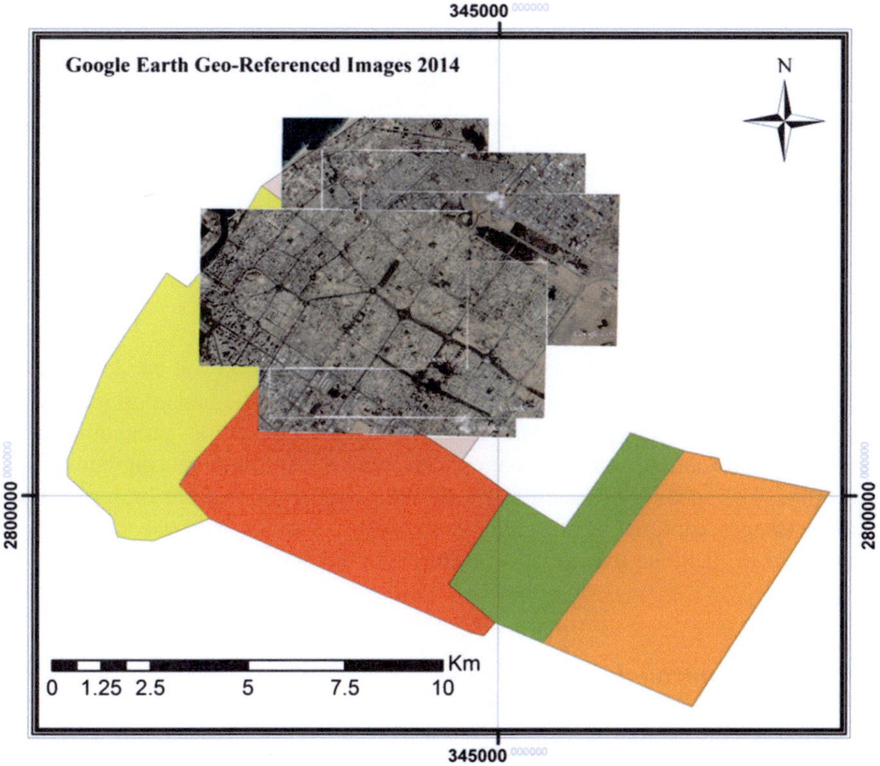


**Fig. 3** 2014 Google earth geo-referenced images for study area

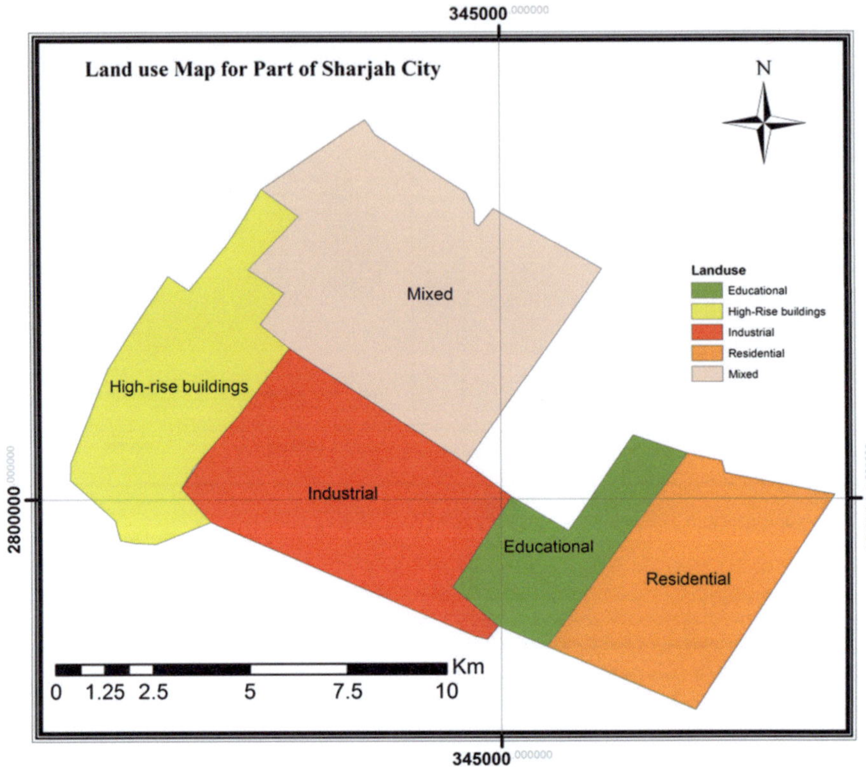


**Fig. 4** Land use data for the study area

# 4 Analysis

In this study, two basic approaches were applied and followed to detect urban growth and change for the study area within Sharjah City. The first approach rely on image processing techniques and accurate image registration procedure using multi-temporal Landsat satellite images with ground resolution varying between 15 and 60 m. The second approach relies on digitizing features from high-resolution Google images captured at different years, where various land-use classes were established to facilitate the growth development analyses.

## *4.1 First Approach*

Figure 5a shows the resampled images after applying previous transformation where corresponding pixels are pointing at the same object space feature. It has to

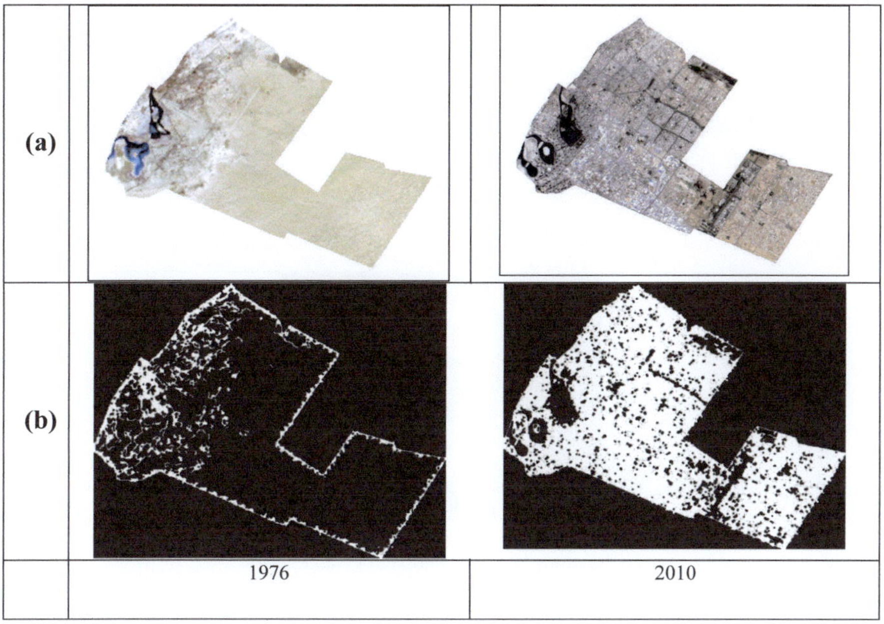


**Fig. 5** **a** Cropped Landsat image for years 1976 and 2000. **b** Derived edges

be mentioned that radiometric differences and noise among images were reduced by applying various image processing techniques (Al-Ruzouq and Habib 2012; Cavallaro and Touradj 2001; Agouris et al. 2000; Canny 1986). Figure 5b shows a sample of derived edges for the study area at different years where a simple pixel-by-pixel subtraction between the resampled edge images was used to extract changing pixels. The filtered images emphasized areas with interesting features since they would lead to a dense distribution of edge cells. Afterward, the filtered images were subtracted to highlight areas of change. Figure 6 shows the difference image from 1987 until 2010. In the difference image, white areas indicate changes while black areas indicate parts with no change. From the visual inspection of Fig. 6, it can be clearly noticed that areas where change have occurred are represented by white pixels. AL-Ruzouq and Shanableh (2014) illustrated that the highest range of growth represented by linear features (building and roads) was accrued between 1976 and 1987, which consists of 36.24% of the total urban features in Sharjah City. Moreover, the result showed that between 1976 and 2010, the cumulative urban expansion inside Sharjah City is 71.9% (Al-Ruzouq and Shanableh 2014).

**Fig. 6** Difference image for the study area for years 1987 and 2010

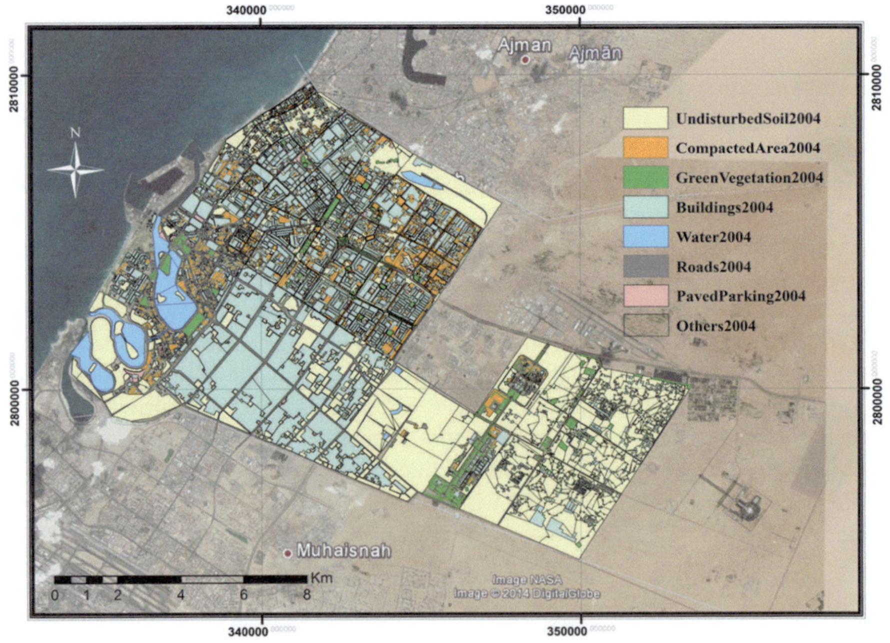


**Fig. 7** Digitized features from 2004 geo-referenced Google earth image

## 4.2 Second Approach

In this stage, multi-temporal Google Earth images were used to digitize various features. The features were classified into the following nine different classes: Roads, Buildings, Paved Parking, Compacted Soil, Undisturbed Soil, Green Area, Water, and Others. Figure 7 shows the digitized features from the 2004 geo-referenced Google Earth image.

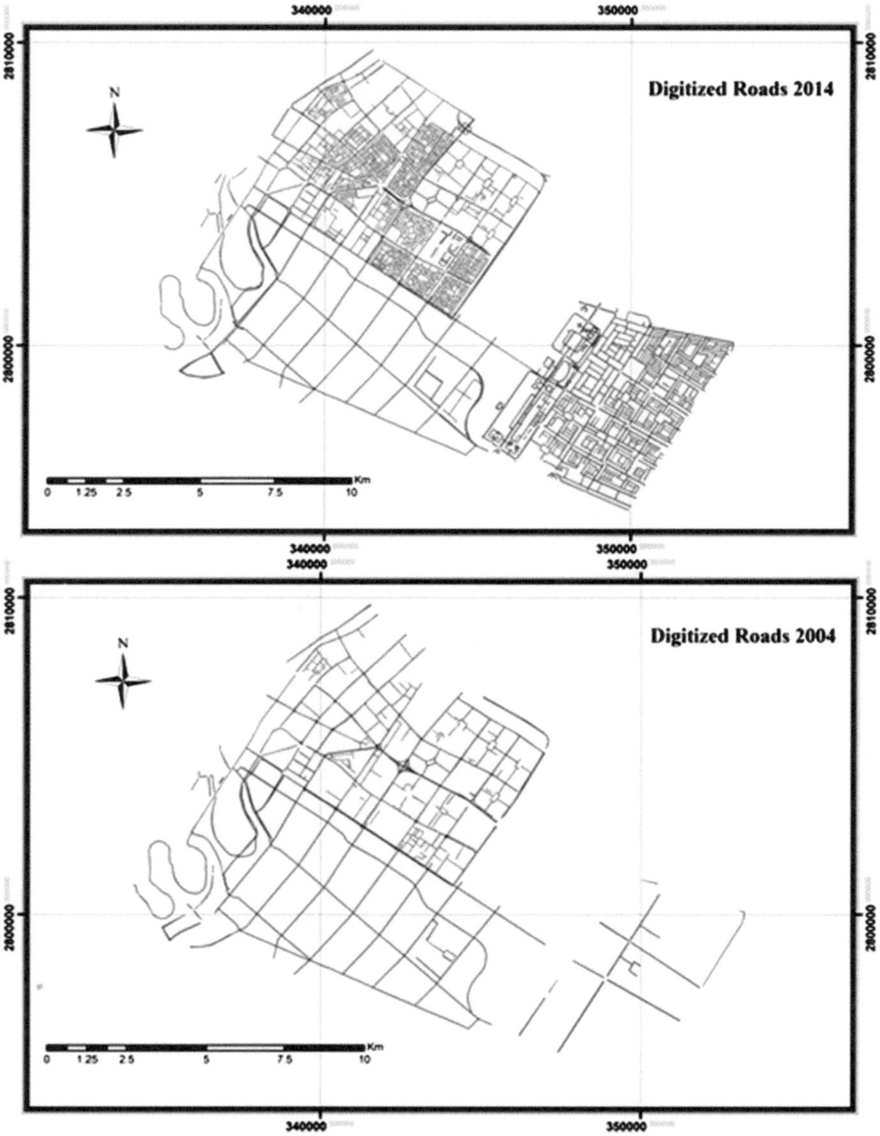


**Fig. 8** Digitized roads from geo-referenced Google earth image for years 2014 and 2004

Figure 8 shows the digitized roads for the study area for the years 2004 and 2014. The expansion of the road and street network can be easily visualized in the newly developed parts of the study area at the lower right corner. Simple calculation of polygons areas for linear features of roads and buildings show that the range of growth represented by linear features (building and roads) that occurred during the 1976–2014 period accounts for about 33% of the total area inside Sharjah City.

## 5 Conclusion

This paper presents an image registration together with a suggested procedure for detecting urban development changes from imagery. The proposed approach has been tested on a real-world setting, which showed its effectiveness in registering and detecting changes among multi-temporal and multi-resolution imagery. Digitized features based on geo-referenced Google Earth images were extracted to work as ground truth for the study area. Future research will concentrate on using high-resolution images for change detection and simultaneously establishing ground truth for quantitative evaluation of the proposed approach.

## References

Agouris P, Mountrakis G, Stefanidis A (2000) Automated spatiotemporal change detection in digital aerial imagery. In: SPIE proceedings 4054:2–12

Al-Ruzouq R, Habib A (2012) Linear features for automatic registration and reliable change detection of multi-source imagery. J Spat Sci 57(1):51–64 (Taylor & Francis)

Al-Ruzouq R, Shanableh A (2014) Multi-temporal satellite imagery for urban expansion assessment at Sharjah city/UAE. Malays J Remote Sens GIS 3(2):76–83

Bruzzone L, Prieto D (2000) Automatic analysis of the difference image for unsupervised change detection. IEEE Trans Geosci Remote Sens 38(3):1171–1182

Canny J (1986) A computational approach to edge detection. IEEE Trans Pattern Anal Mach Intell 6(6):679–698

Cavallaro A, Touradj E (2001) Change detection based on color edges. In: Proceedings of IEEE international symposium on circuits and systems (ISCAS-2001)

Dowman I (1998) Automated procedures for integration of satellite images and map data for change detection: the archangel project. GIS-between visions and applications. IAPRS 32 (4):162–169

Habib A, Al-Ruzouq R (2004). Line-based modified iterated hough transformation for automatic registration of multi-source imagery. Photogram Rec 19(105):5–21

Habib A, Kelley D (2001a) Automatic relative orientation of large scale imagery over urban areas using modified iterated hough transform. Int J Photogram Remote Sens 56(2001):29–41

Habib A, Kelley D (2001b) Single photo resection using the modified hough transform. J Photogram Eng Remote Sens 67(2001):909–914

Singh A (1989) Review article digital change detection techniques using remotely-sensed data. Int J Remote Sens 10(6):989–1003

Wikipedia (2016) Demographics of the United Arab Emirates. Available at: https://en.wikipedia.org/wiki/Demographics_of_the_United_Arab_Emirates (online). Accessed 21 Sept 2015

# Part IV
# Climate Change

# Assessment of the Potential Impacts of Sea Level Rise on the Coastal Plain of Al Batinah, Sultanate of Oman

**Salim Mubarak Al Hatrushi**

**Abstract** The fourth assessment report of the Intergovernmental Panel on Climate Change concluded that sea level rise (SLR) will range between 18+ centimeters and 59+ cm by 2100. Furthermore, the report indicates that it is most likely SLR will be up to 100+ cm. The Coastal Plain of Al Batinah extends for more than 200 km, accommodates about 28% of total population in Oman, and considered the economic hub of Oman.This paper aims to determine the potential areas that will be affected by SLR using GIS and remote sensing techniques. The investigation is based on IKONOS Images (1 m), digital elevation model (DEM 30 m), field work observation and data base of tide level from the Hydrographic National Bureau of Oman. According to the following scenarios, SLR inundation level of 1m will result in a shoreline retreat of about 3400 m. SLR inundation level of about 2 m will result in a shoreline retreat of about 3700 m. The worst SLR inundation level of 5 m will result in a shoreline retreat of about 5100 m. El-Sawadi coast will be the most affected area.

**Keywords** Sea level rise · IPCC scenarios · Al Batinah · Oman · Inundation level

## 1 Introduction

Coastal geomorphic processes of erosion and accretion and sea level changes continuously modify the shoreline. The primary causes of coastal erosion and landward retreat of the shoreline are direct wave action, the interception of littoral drift, sea level rise, and human interference. The sea level rise at any particular location is controlled by processes such as melting of polar ice caps and glaciers, changes in sea temperature and thermal expansion, seafloor spreading and undersea volcanism, changes due to localized subsidence or uplift of the coastline, and

S.M. Al Hatrushi (✉)
Department of Geography, College of Arts and Social Sciences,
Sultan Qaboos University, Muscat, Oman
e-mail: hatrushi@squ.edu.om



S. Pirasteh and J. Li (eds.), *Global Changes and Natural Disaster Management: Geo-information Technologies*, DOI 10.1007/978-3-319-51844-2_10

changes from the deposition of sediment. Ocean thermal expansion and high waves associated with extreme meteorological events (atmospheric depressions) can also cause temporary coastal erosion. The fourth assessment report of the Intergovernmental Panel on Climate Change (IPCC) concluded that sea level rise (SLR) will range between +18 and +59 cm by 2100. Furthermore, the report indicates that it is most likely SLR will be up to +100 cm. Al Batinah Coastal Plain is heavily settled, and it is one of the most fertile areas in Oman. It accommodates about 28% of total population in Oman and is considered the economic hub of Oman.

Studies on SLR and coastal erosion and landward retreat of the shoreline in Oman are limited and mainly investigating coastal erosion in Al Batinah. Dobbin (1992) reported an extensive study on the status of erosion along the coasts of Oman. They concluded that some coastal areas in Oman are experiencing severe erosion due to the reduction in sediment supply by recharge dams and roads, construction of harbors and the long-term process of sea level rise. Badenhorst (1995) considered the Al Batinah coast to be in a state of widespread erosion with recharge dams, sea level rise, and harbor construction all being possible causes. In their Environmental Impact Assessment, Atkins (2002) investigated the coastal area adjacent to Sohar Industrial Port. They have related the existed coastal erosion problem mainly to the coastal engineering activities such as harbor development. Al-Hatrushi et al. (2014) investigated the status of coastal erosion in Al Batinah and suggested that Al Batinah sandy shoreline is stable. However, local interruption of longshore transport by coastal engineering structures has caused limited erosion and accretion on short temporal and spatial scales. Erosion related to natural processes such as sea level rise was not investigated in this study.

This paper aims to determine the potential impacts of sea level rise on the Coastal Plain of Al Batinah and assesses the potential areas that will be affected by SLR, using scenarios based on the following expected sea level rise of 1, 2, 3, 4, and 5 m.

## 2 Study Area

The Batinah coastal plain is located in the northeastern part of the Sultanate of Oman (Fig. 1), bounded between the Western Hajar Mountains and the Sea of Oman. It extends in an NW–SE trend as a crescent shape and parallel to the mountains for about 230 km from UAE border in the NW to Ras Al Hamra in Muscat in the SE. The Batinah coast is mostly sandy and characterized by longshore sand transport along a low-lying coastal plain fringed by beaches and dunes. The coastal plain is narrow at its northern and eastern ends while its width in the middle is maximum around 50 km. The coastal plain is composed of continuous alluvial fans drained from the mountains with sediment which varies from gravel and coarse sands to fine sands and silt near the coast. This suggests that most of the beach sediments along the coast are of terrestrial origin.

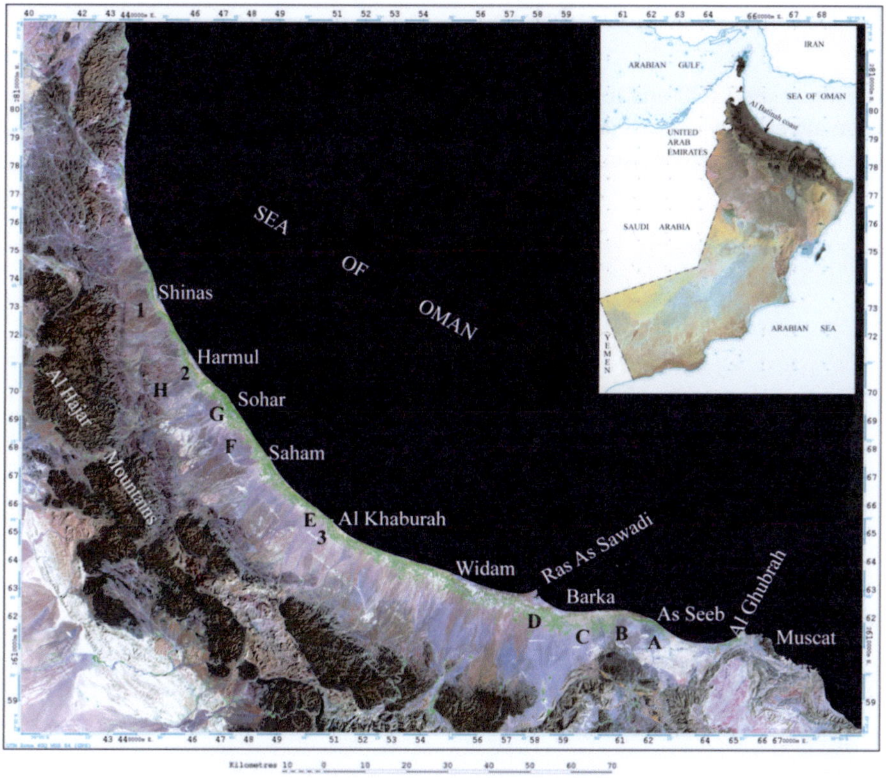


**Fig. 1** The study area

According to the census of 2010, 80% of the Omani population lives in low-lying areas such as coastal plains and could be threatened by sea level rise. During the last five years, the country witnessed two strong tropical cyclones, Gonu 2007 and Phet 2010 which cause a damage estimated at 5 billion US$ and revealed the extreme vulnerability of the populated Omani coast to storm surge and flooding (Charabi 2010).

Human settlement along the Batinah coast follows the pattern of an almost continuous strip of towns and villages starting from Seeb in the southeast to Shinas in the northwest. Residential houses and commercial buildings are built close to or on the beach (Dobbin 1992). For centuries, houses used to be fabricated from only semi-permanent materials such as palm fronds. Today all the building are constructed from solid materials.

## 3 Dataset and Methodology

The study used GIS and remote sensing techniques to determine the areas expected to be affected by expected SLR, and to map the areas affected and finally to make suggestions for the decision makers regarding this issue. The investigation is based

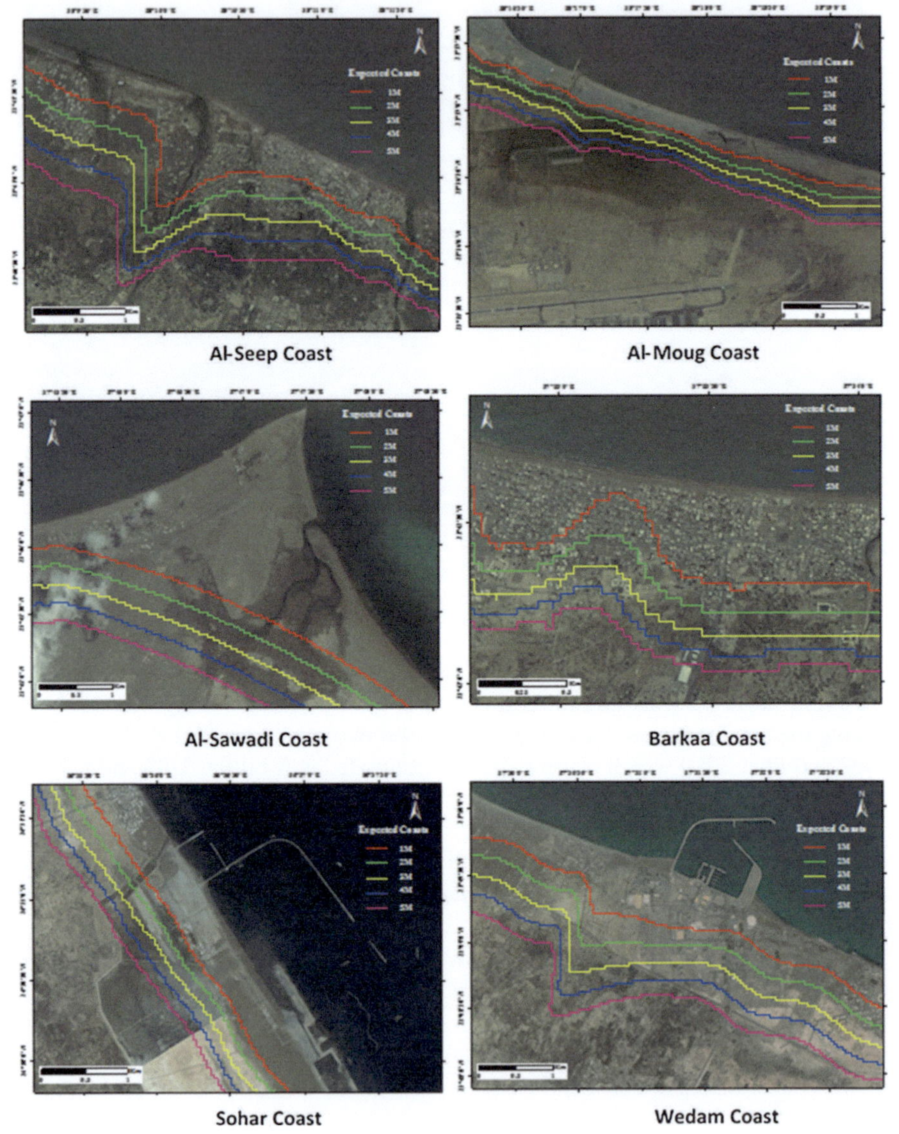


**Fig. 2** The possible affected for climate changes on the coasts of main cities in Al Batinah until Contour +5 M

**Table 1** Expected coastline changes (m) for main cities in Al Batinah during sea level rise

| City | Con. 1 | Con. 2 | Con. 3 | Con. 4 | Con. 5 | Max. |
|---|---|---|---|---|---|---|
| Al Moug | 425 | 515 | 631 | 820 | 1012 | 2110 |
| Al Seep | 705 | 920 | 1105 | 1340 | 1910 | 3450 |
| Barkaa | 607 | 806 | 1075 | 1112 | 1350 | 1507 |
| Al Sawady | 3410 | 3720 | 4100 | 4320 | 4710 | 5100 |
| Wedam | 605 | 816 | 1020 | 1450 | 1920 | 2410 |
| Sohar | 420 | 514 | 621 | 715 | 820 | 975 |

on IKONOS Images (1 m), digital elevation model (DEM 30 m), field work, and database of tide level from the Hydrographic National Bureau of Oman.

## 4 Results and Discussions

It is expected that Al Sawadi coastline and the lagoon associated with a surface covered mainly with fine sand and silt will witness the most potential impacts of sea level rise on the Coastal Plain of Al Batinah (Fig. 2c). According to the scenarios of SLR inundation level of 1 m will result in a shoreline retreat of about 3400 m while SLR inundation level of about 2 m will lead to a shoreline retreat at Al Sawadi coastline of about 3700 m. The worst scenarios of SLR inundation level of 5 m will result in a shoreline retreat of about 5100 m. NW Al Seeb (Fig. 2a) will be the second most potential impacts of sea level rise on the Coastal Plain of Al Batinah according to these scenarios, where the shoreline will retreat for about 705 m with an SLR scenario of 1 m (Table 1).

To conclude, due to its low-lying nature, Al Batinah coastline has a great potential to the impacts of sea level rise. Furthermore, there are certain locations which affected more such as Al Sawady coastline and Al Seeb coastline. To protect social communities, setback requirement should be implemented on any engineering structure planned along the coastline.

## References

Al-Hatrushi SM, Kwarteng AY, McLachlan A, Sana A, Al-Buloushi AS, Hamed K (2014) Coastal erosion in Al Batinah, Sultanate of Oman. Academic Publication Board, Sultan Qaboos University, Muscat, p 261

Atkins (2002) Shoreline evolution report. Sohar Port project, Batinah coast Oman. Final Report, Muscat, Oman

Badenhorst P (1995) Erosion along the Batinah coastline, Gulf Petrochemical Services. CSIR/GPS Report, a short report on-site visit

Charabi Y (2010) Indian Ocean tropical cyclones and climate change. Springer Publications, India, p 371

Dobbin J (1992) Coastal erosion in Oman: draft regulations for the prevention of coastal erosion in the Sultanate of Oman. Final draft report, May 1992 to Ministry of Regional Municipalities and Environment, Muscat, Oman

# Climate Change and Insecurity: An Examination of Gombe State's Predicament in the Northeastern Nigeria

**U.A. Abubakar and A. Ahmed**

**Abstract** Insecurity as a result of climate change has become a global phenomenon, which imposes a major challenge to human environments especially in the twenty-first century. This phenomenon prompted scholars to conduct scientific investigations to find out solutions to the problems. The overwhelming challenges related to socioeconomic activities, migration, and agricultural productions eventually resulted in inadequate employment opportunities and limited means of the lively hood to the teaming population. Accordingly, it intensifies social vices which culminate into general insecurity causing loss of lives and valuable properties. This study investigated climate change pertaining insecurity in Gombe state within North East. Three research questions guide the study, and one null hypothesis was tested at 0.05 level of significance. The descriptive research design of survey type was employed. The research questions raised were answered using mean and standard deviation, while research hypothesis rose was verified using *t*-test with the aid of SPSS. A validated instrument with a reliability coefficient of 0.84 guided the collection of data. The study observed that climate change is a major factor responsible for insecurity in Gombe state within the region which results in desertification, a decrease in water bodies, and loss of soil fertility, extinction of flora and fauna, and an increase of temperature. The study recommended that Nation states, civil society organizations, and individuals should respect international laws, legislations, treaties, policies, programs, and resolutions regarding our environments, and finally individuals should also endeavor to imbibe the culture of a friendly environment.

**Keywords** Insecurity · Climate · Change · Environment · Development

U.A. Abubakar (✉) · A. Ahmed
Department of Political Science, Federal University Kashere, Gombe State, Nigeria
e-mail: umaralhaji12@gmail.com; umaralhaji@fukashere.edu.ng

A. Ahmed
e-mail: eswhy92@yahoo.com



S. Pirasteh and J. Li (eds.), *Global Changes and Natural Disaster Management: Geo-information Technologies*, DOI 10.1007/978-3-319-51844-2_11

## 1 Background of the Study

The phenomenon of climate change over the years has remained an issue of concern amongst scholars and policy makers in the world and Nigeria in particular. This has prompted scientific investigations to find out solutions to the existing problems. Climate change has its relates to human security on (1) economic activities such as shortage of food and cash crops productions, widespread of poverty, and income inequality; (2) migration issues especially on conflict between pastoralists and the farmers, settlers, and the indigenes and other communal and land related disputes; (3) agricultural production particularly on drought, flooding, erosion and heavy storm that affect food production and lead to hunger and starvation; and others. Consequently, these situations exacerbated social vices and this culminated into the general insecurity that caused the loss of lives and valuable properties not only in northeastern Nigeria but also in Africa and the world at large.

Although, there are numerous studies conducted on climate change related to its impacts on health, food supply, economic growth, migration, societies, and public properties, (such as water, land), and some changes related to global warming. However, less attention is given to effects of climate change on human security that lead to many forms of violence specifically within the northeastern Nigeria.

## 2 Review of Literature on Climate and Insecurity

The 2011 Report by the American Climate Choice stated that climate change has brought about severe and possibly permanent alterations to our planet's geological, biological and ecological systems. The report suggested that our natural environment is affected by climate change, and that has long-term effects on the survival of human and nonhuman species. Similarly, McMichael (2003) also stated that there is new and strong evidence that most of the warming observed over the last 50 years is attributable to human activities within our environment. This explanation linked climate change to human security on factors, such as industrialization urbanization and technological advancement in the twentieth and twenty-first centuries. The factors mentioned above as furthermore explained by McMichael et al. (2006), Sahney et al. (2010), and MPIBGC/PH (2013) have led to the emergence of large-scale environmental hazards to human health, such as weather, ozone layer depletion increased danger of wildland fires, loss of biodiversity, stresses to food-producing systems, and the global spread of infectious diseases.

Accordingly, Lunde and Lindtjorn (2013) and MPIBGC/PH (2013) have stressed that majority of the adverse effects of climate change are experienced by poor and low-income communities around the world, who have much higher levels of vulnerability to environmental determinants of health, wealth, and other factors, and much lower levels of capacity available for coping with environmental change. A report by Yale Environment on the Global Human Impact of Climate Change published by the Global Humanitarian Forum in 2009, estimated that more than 300,000 deaths and about $125 billion in economic losses each year, and this indicated that most climate change induced mortality as due to worsening floods and droughts in developing countries. It means that Gombe state of northeastern Nigeria is one of the areas most affected by the devastating effects of climate change in Nigeria.

## 3 Impacts of Climate Change on Human Security

Climate change has, directly and indirectly, affected the security policies of nation states, regional bodies and the global systems through global warming. Harrod and Martin (2013) acknowledged that, though, climate change does not always lead to violence and conflicts but are often caused by multiple interconnected factors. However, Mimura et al. (2007) and Patwardhan et al. (2007) observed that climate change has a potential to exacerbate the existing tensions or create new ones and served as a threat multiplier. It can also be a catalyst for violence and conflict and it may foster a threat to international security. Accordingly, Wilbanks et al. (2007) added that the IPCC has suggested that the disruption of internal and external migration may serve to exacerbate conflicts, though they are less confident of the role of increased resource scarcity.

Furthermore, varieties of experts have warned that climate change may lead to high conflict. For instance, the US Military Advisory Board (2007) argued that there be a fear that global warming will serve as a "threat multiplier" in the volatile regions. Furthermore, the Center for Strategic and International Studies and the Center for a New American Security, two Washington think tanks, have reported that flooding "has the potential to challenge regional and even national identities," leading to "armed conflict over resources" (Campbell et al. 2007). They indicate that the greatest threat would come from large-scale migrations of people both inside nations and across existing national borders. The nexus between climate change and human security is a concern of scholars and policy makers across the globe. The impacts of climate change on human security can be seen as a threat which is expected to hit developing nations. In developed countries, "an unstable climate will exacerbate some of the core drivers of conflict, such as migratory pressures and competition for resources" (Campbell et al. 2007).

Scholars argued for connections between climate change and insecurity in their studies of some conflicts in Africa and other continents of the world. For instance, the Somali Civil War is related to drought and extreme temperature that fueled the violence; the War in Darfur in Southern Sudan is also due to the drought that triggers the conflict between herders and farmers. The Syrian Civil War is also fueled by the element of drought that caused very low production capacity of agricultural products which led to death and displacement of thousands and millions of people respectively. The Nigerian phenomenon of Boko Haram Insurgency in northeastern Nigeria region is also due to the partly drying-up of Lake Chad desertification that subsequently led to the loss of jobs to youths farmers, pastoralists, and fishers. The Niger Delta militancy in the southern region is also related to activities of Multinational Companies that polluted the natural environment which affected agricultural activities especially fishing providing the teaming population with jobs in the region. It is against this background that this study is conducted to explain the impacts of climate change on human security in Gombe state of northeastern Nigeria.

## 4 Objectives of the Study

The study has the following objectives:

1. To determine the forms of climate change as it affects human security in Gombe state of northeastern Nigeria.
2. To examine the effects of climate change on human security in Gombe state of northeastern Nigeria.
3. To suggest different strategies or approaches to reducing the effects of climate change on human security in Gombe state of northeastern Nigeria.

## 5 Research Questions

To achieve these, the study developed three research questions and two null hypotheses.

1. What are the forms of climate change on the security situation in Gombe state of northeastern Nigeria?
2. How climate change impacted on security situations of both rural and urban areas of Gombe state in northeastern Nigeria?
3. What are the strategies and approaches to be adopted to reduce the impacts of climate change on human security in Gombe state of northeastern Nigeria?

## 6 Null Hypothesis

The study tested one null hypothesis at 0.05 level of Significance.

**Ho**: There is no significant difference in the perception of respondents regarding the effect of climate change on security challenges in rural and urban areas of Gombe state in northeastern Nigeria.

## 7 Methodology

The study utilized descriptive research design of survey type. Samples of 2500 respondents were randomly selected from six rural and urban areas of Gombe state, northeastern Nigeria. The researchers utilized the instrument and tagged it as "Vulnerability to Climate Change and Security Challenges Questionnaire (VCCSCQ)". This is validated by three experts of rank senior lecturers and above with Ph.D. qualification, and reliability coefficient of 0.84 guided in the collection of data. Focus group using VCCSCQ helped in gathering data from the respondents. Mean and standard deviations were employed in answering research questions, while *t*-test statistics at 0.05 level of significant aid in testing null hypothesis rose. Statistical analysis involved the use of computer software called SPSS Package.

## 8 Presentation of Findings

1. What are the forms of climate change on security situations in Gombe state of northeastern Nigeria?

Table 1 presents information on forms of climate change and how it affects security situation in Gombe state of northeastern Nigeria. From the table, Erosion attracted a mean of 65.1, desertification 54.2, flood, 76.5, drought 58.2, global warming 12.1, cyclones 10.2, storm 12.2, and extreme weather 43.6. From the result, the flood has the highest mean, followed by erosion, drought, and desertification.

2. How climate change impacted on security situations of both rural and urban areas of Gombe state in northeastern Nigeria?

Table 2 presents the perception of respondents in rural and urban areas regarding the effect of climate change on security situations. From the information obtained, the mean and standard deviation of respondents in the rural and urban areas shows the following: hunger (12.1; 43.6); Poverty (10.2; 43.6); adverse health outcomes of

**Table 1** Mean and standard deviation of respondents' opinion on how climate change affect security situation in Gombe state of northeastern Nigeria

| Aspect of climate change | Mean | s.d. |
|---|---|---|
| Erosion | 65.1 | 21.1 |
| Desertification | 54.2 | 15.7 |
| Flood | 76.5 | 23.2 |
| Drought | 58.2 | 13.2 |
| Global warming | 12.1 | 4.5 |
| Cyclones | 10.2 | 5.2 |
| Storm | 12.2 | 7.1 |
| Extreme weather | 43.6 | 14.2 |

**Table 2** Perception of respondents in rural and urban areas regarding the effects of climate change on security situations

| Effects | Rural | | Urban | |
|---|---|---|---|---|
| | Mean | s.d. | Mean | s.d. |
| Hunger | 12.1 | 4.5 | 43.6 | 14.2 |
| Poverty | 10.2 | 5.2 | 43.6 | 14.2 |
| Adverse health outcomes and spread of infectious diseases | 39.3 | 12.1 | 38.3 | 11.1 |
| Competition for resources | 45.8 | 21.2 | 50.8 | 11.2 |
| Loss of habitat | 37.2 | 13.5 | 39.2 | 8.5 |
| Threat of violence and armed conflict | 56.3 | 11.2 | 62.3 | 12.2 |
| Displacement and migratory pressures | 54.5 | 8.5 | 49.5 | 8.5 |
| Increased danger of wild land fires | 45.8 | 21.2 | 50.8 | 11.2 |
| Stresses to food-producing systems | 54.5 | 8.5 | 49.5 | 8.5 |

malnutrition, diarrhea, injuries, cardiovascular and respiratory diseases, and water-borne diseases and insect-transmitted diseases (39.3; 38.3); competition for resources (45.8; 50.8); habitat loss (37.2; 39.2); threat of violence and armed conflict (56.3; 62.3); displacement and migratory pressures (54.5; 49.5); increased danger of wildland fires (45.8; 50.8); and stresses to food-producing systems (54.5; 49.5).

3. What are the strategies/approaches to climate change as it relates to security issues adopted by people in Gombe state of northeastern Nigeria?

Table 3 presented the strategies/approaches to climate change as it relates to security issues adopted by respondents in Gombe state of northeastern Nigeria. The result showed that afforestation, reduction of fuel utilization, sensitization by community members, legislation drainage construction, collaboration with neighboring states/countries on security and environmental issues, awareness campaign, and sensitization on the importance of preserving our environment by civil society organizations, imbibing the culture of friendly environment by individuals, control grazing, use of biofuels, and use of law enforcement agencies are the strategies adopted by the respondents in respond to climate change.

**Ho**: There is no significant difference in the perception of respondents regarding the effect of climate change on security challenges in rural and urban areas of Gombe state in northeastern Nigeria.

Table 4 presents *t*-test analysis of respondents' perception regarding the effect of climate change on security challenges in rural and urban areas of Gombe state in northeastern Nigeria. From the table, *t*-obtained is 1.98 and *p* observed is 0.08 at 0.05 level of significance and degree of freedom of 2498. Since the *p* value observed is greater than alpha value, there is no significant difference. This implies retaining of null hypothesis which stated that there is no significant difference in the perception of respondents regarding the effect of climate change on security challenges in rural and urban areas of Gombe state in northeastern Nigeria.

**Table 3** Mean and standard deviation of respondents opinion on strategies/approaches adapted to climate change as it relates to security situations

| Approach | Mean | s.d. |
|---|---|---|
| Afforestation | 51.8 | 11.2 |
| Sensitization by community members | 37.2 | 8.5 |
| Reduction of fuel utilization | 22.3 | 12.2 |
| Legislation | 49.5 | 8.5 |
| Control of bush burning | 50.8 | 11.2 |
| Drainage construction | 39.2 | 8.5 |
| Collaboration with neighboring states/countries on security and environmental issues | 13.3 | 2.2 |
| Awareness campaign and sensitization on the importance of persevering our environment by civil society organizations | 19.5 | 8.5 |
| Imbibing the culture of friendly environment by individuals | 10.8 | 1.2 |
| Control grazing | 36.5 | 8.5 |
| Use of biofuels | 12.3 | 2.2 |
| Use of law enforcement agencies | 39.5 | 8.5 |

**Table 4** The *t*-test analysis of respondents through perception on the significant difference in both rural and urban areas of Gombe state

| Variable | *N* | Mean | s.d. | d.f. | *T* | Alpha | *p* | Remark |
|---|---|---|---|---|---|---|---|---|
| Rural | 1150 | 68.9 | 12.2 | 2498 | 1.98 | 0.05 | 0.08 | Not significant |
| Urban | 1350 | 69.2 | 13.5 | | | | | |

# 9 Discussion of Findings

First, result presented in Table 1 revealed the relationship between climate change and insecurity and how it affects human and social security in Gombe state of northeastern Nigeria. Erosion desertification, flood, drought, global warming, cyclones, storm, and extreme weather are the major attributes of climate change that inversely affected security situation in Gombe state. Thus, the result shows that flood has the highest mean, followed by erosion, drought, and desertification.

Second, Table 2 unveiled the views of respondents in rural and urban areas regarding the effects of climate change on security situations. The results show that hunger, poverty, adverse health outcomes from malnutrition, diarrhea, injuries, cardiovascular, and respiratory diseases are some of the effect of climate change on human security. Other effects included water-borne and insect-transmitted diseases; competition for limited resources, loss of habitat, the threat of violence, armed conflict, displacement and migratory pressures, increased the danger of wildland fires and stresses to food-producing systems constituted social implications of climate change on the populace of Gombe state.

Thirdly, Table 3 described the strategies/approaches to overcome the major challenges and effects of climate change to improve the security wise of all aspects of human endeavor. The result showed that afforestation, reduction of fuel utilization, sensitization of community members, legislation on drainage construction, collaboration with neighboring states/countries on security, and environmental issues. Other important ways included awareness campaign and sensitization on the importance of preserving our environment by civil society organizations, imbibing the culture of the friendly environment by individuals, control grazing, use of biofuels, and use of law enforcement agencies are the strategies that will address the effects of climate change on human and environmental security in Gombe state, northeastern Nigeria and the country at large.

Finally, Table 4 presented $t$-test analysis of respondents' views on the effects of climate change on security in both rural and urban areas of Gombe state. Therefore, the result revealed that $t$-obtained is 1.98 and $p$ observed is 0.08 at 0.05 level of significance and degree of freedom of 2498. Since the $p$ value observed is greater than alpha value, there is no significant difference. Thus, the finding implies retaining of null hypothesis which stated that there is no significant difference in the perception of respondents regarding the effect of climate change on security challenges in both rural and urban areas of Gombe state.

The findings are in line with that of previous studies of McMichael et al. (2006), Sahney et al. (2010), and MPIBGC/PH (2013) ascertained that, climate change has brought about severe and possibly permanent alterations to our planet's geological, biological and ecological system, emergence of large-scale environmental hazards to human health, such as weather, ozone, increased danger of wildland fires, loss of biodiversity, stresses to food production and the global spread of infectious diseases, food supply, economic growth, migration security, societal change, and public properties, such as water, land, and changes related to global warming. Similarly, Mimura et al. (2007) and Patwardhan et al. (2007) also argued that climate change has the potential to exacerbate existing tensions or create new ones serving as a threat multiplier. It can be a catalyst for violent conflict and a threat to international security.

## 10 Conclusion

The findings emanated from this study revealed that climate change has the potential to exacerbate tension in Gombe state of northeastern Nigeria. The study observed that climate change is a major factor responsible for insecurity in the region which results in desertification, a decrease in water bodies, and loss of soil fertility, extinction of flora and fauna and an increase of temperature. This is evident in a focus group discussion with the respondents who claimed that climate change, such as flooding has destructed many houses, properties, and valuables, lead to death and induced migration. Thus, it becomes a phenomenon that required attention of agencies and asunder globally.

## 11 Recommendations

This research has the following recommendations among others:

1. There is a need for nation states to respect international resolutions on environmental protection.
2. There is a need for federal and state governments in Nigeria to strengthen the application of existing legislations on the environment.
3. There is also a need for federal and state governments in Nigeria to strengthen its collaboration with neighboring countries on security and environmental issues.
4. Civil society organizations should strengthen the awareness campaign and sensitization programs on the importance of preserving our natural environment.
5. Individual and members of civil societies should endeavor to imbibe the culture of a friendly environment.

## References

Campbell KM, Gulledge J, McNeill JR, Podesta J, Ogden P, Fuerth L, Woolsey RJ, Lennon AT, Smith J, Weitz R, Mix D (2007) The age of consequences: the foreign policy and national security implications of global climate change. Center for Strategic and International Studies, Washington DC

Harrod RP, Martin DL (2013) Bioarchaeology of climate change and violence: ethical considerations. Springer Science & Business Media

Lunde TM, Lindtjørn B (2013) Cattle and climate in Africa: how climate variability has influenced National cattle holdings from 1961–2008. Peer J 1(1):e55. doi:10.7717/peerj.55. www.pdfsearchengine.org

McMichael AJ (2003) Global climate change and health: an old story at large. In: McMichael A, Campbell-Lendrum D, Corvalan C, Ebi K, Githeko A, Scheraga J, Woodward A (eds) World Health Organization, Geneva

McMichael AJ, Woodruff R, Hales S (2006) Climate change and human health: present and future risks. Lancet 367(9513):859–869

Mimura N, Nurse L, McLean RF, Agard J, Briguglio L, Lefale P, Payet R, Sem G (2007) Small islands. Clim Change 687–716

MPIBGC/PH (2013) Extreme meteorological events and global warming: a vicious cycle? Max Planck Res

Munich Climate-Insurance Initiative (2013) Climate change and rising weather related disasters. www.pdfsearchengine.org

Patwardhan A, Semenov S, Schnieder S, Burton I, Magadza C, Oppenheimer M, Pittock B, Rahman A, Smith J, Suarez A, Sukumar R (2007) Assessing key vulnerabilities and the risk from climate change. Clim Change 779–810

Sahney S, Benton MJ, Ferry PA (2010) Links between global taxonomic diversity, ecological diversity and the expansion of vertebrates on land (PDF). Biol Lett 6(4):544–547. www.pdfsearchengine.org

Wilbanks TJ, Lankao PR, Bao M, Berkhout F, Cairncross S, Ceron JP, Kapshe M, Muir-Wood R, Zapata-Marti R (2007) Industry, settlement and society. Clim Change 357–390

# Climate Change and Forced Migration from Ngala and Kala-Balge LGAs, N.E. Borno State, Nigeria

**Adam M. Abbas**

**Abstract** North-Eastern Borno State, Nigeria, due to its location, size, and population is very vulnerable to the impact of climate change. Even though sufficient understanding and propagation are made on the impact of climate change and its associated problems, little effort is however made to address most of the problems emanating from it. More especially forced migration, one of the resultant effects of drought, a component of climate change is given less attention in this part of the country. Specifically, the paper explores how the expected high intensity of droughts in the study area might worsen crop production situation and lead to the only adaptation strategy, a widespread forced migration. This paper focuses on the climate change impact and one of the effects, migration, and its associated problems. Purposive sampling technique was adopted in sampling 250 respondents who were mainly family members of out-migrants from Ngala and Kala-Balge LGAs, Borno State, Nigeria. Available literature was also consulted for the types of climate change impacts. The results revealed that climate change leads to climatic variation over the space with numerous effects on the environment such as intermittent droughts, desertification/deforestation, low water table, and the establishment of dams upstream across the courses of the main sources of water supply to the Lake Chad hence, low agricultural production especially rain-fed. Many people in the study area either migrated to Cameroon's Darrak, Bullaram, Lake Doi, Lake Chad, and Mayo-Mbund for fishing and petty trading or South-Western Nigeria especially Lagos, Oyo states, etc., to serve as security guards and other low-skilled workers, leaving all or some members of their families at home. More than half of respondents (58%) indicated that the head of the households migrated as a result of poor harvest due to diminishing or fluctuating rains/drought and/or drying of river Surbewel. It is recommended that interbasin water transfers should be embarked upon.

**Keywords** Migration · Climate · Drought · Dam · Sudan-Sahel · Intermittent

A.M. Abbas (✉)
Department of Geography, Federal University Kashere, Gombe-State, Nigeria
e-mail: dradamabbas28@gmail.com



S. Pirasteh and J. Li (eds.), *Global Changes and Natural Disaster Management: Geo-information Technologies*, DOI 10.1007/978-3-319-51844-2_12

# 1 Introduction

Climates have changed and still are constantly changing at all scales, from local to global, and over varying time spans. Each "up and down" fluctuation can lead to conditions which are warmer or colder, drier or wetter, stormier or quiescent. Climate change may be due to natural internal processes or external forced, or persistent anthropogenic changes in the composition of the atmosphere or land use (Waugh 1995; Bates et al. 2008; Abaje et al. 2012).

The climate change is driven by a warming earth's atmosphere, the temperature of which is currently rising at a rate of 0.30 °C in a year that will peak in about 2050 at 6.5 °C above the present global average temperatures (IPCC 2007). It encompasses all forms of climatic inconstancy, in other words, it encompasses any differences between long-term statistics of the meteorological elements calculated for different periods but relating to the same area regardless of their statistical nature and physical causes (Akangbe et al. 2015). Mamman (2008) points out that the concern is not only with the natural changes in climate but also on the changes induced by human activities at various levels. This makes climate change one of the greatest environmental problems the world is facing today.

The United Nations Framework on Climate Change defines it as a change of climate which is attributed directly or indirectly to human activities that alter the composition of the global atmosphere and which is, in addition to natural climate variability, observed over a comparable period (Aliyo 2008; Akoteyon and Ogundele 2008; Anyadike 2009). The latest assessment report of the Intergovernmental Panel on Climate Change (IPCC) concluded that most of the observed temperature increases since the middle of the twentieth century could be linked to increasing concentrations of Green House Gases (GHG's) resulting from human activities such as fossil fuel burning and deforestation (Solomon et al. 2007).

The predictions of the effects of climate change touch on every aspect of life support on planet earth (Onakala 2011). According to Kowats et al. (2005), many factors such as physiological adaptation and individual and community wealth will influence both the exposure of individuals and population to climate hazards and their impacts.

The question of environment-induced migration, in particular the idea that climate change would lead to mass migration of populations across long distances has been widely reported.

Demographic research has shown that the most vulnerable households cannot migrate internationally and over long distances and that they are likely to move over short distances only. This phenomenon will likely be reinforced due to a decrease in purchasing power as a consequence of environmental change affecting their livelihoods (Onakala 2011).

## 2 Statement of Research Problem

Historical records indicate that droughts and floods have occurred in this zone frequently in the past. As for droughts, it started from the beginning of the twentieth century, 1903–1905, 1913–1914, 1923–1924, 1931–1932, 1942–1944, 1956, 1972–1973, 1982–1983, and 1986–1987 (Oladipo 1993; Odekunle et al. 2008; Bashir 2008). According to Gazali (2005), even the movements and migrations of Ulama from Kanem Borno into the southern region of Nigeria in fairly large groups during the nineteenth century were caused by political instability, natural calamities like drought, famine, and so on. Several single-case analyses suggest that resource scarcity contributes to movement of people from one place to another. Changing rainfall patterns and more frequent natural disasters arising from anthropogenic climate change will lead to migration.

Long-term rainfall records and other evidence from the Sudano-Sahelian region show that shifting climate dynamics have prevailed in the area for at least the last 10,000 years (Nicholson 1983; Hulme 1992). Virtually every ecological region in the country (from the coast to Sahel), is vulnerable to climate change. It is also fairly well established that one of the most significant climatic variations in the NE region of Nigeria since the late 1960s has been persistent decline in rainfall (Maryah 2004), while Mortimore (1989) observed that all the four types of drought (agricultural, hydrological, meteorological, and ecological) have been experienced in the same region. The manifestations included withering of vegetation and crops, low river discharge, retreating levels of Lake Chad, and death of animals. The problems of drought and desertification have also caused among development planners regarding how to cope with the losses of food production, episodes of food, displacement of population, and declining water and rangeland resources.

Some studies have been carried out in this region in an attempt to understand the rainfall pattern (Maryah and Gadzama 1996; Maryah 2004; Ayoade 1974). However, the results of these studies have shown little about the impact of the changing pattern of migration. Specifically, the paper explores how the high-intensity droughts in the Sudan-Sahel (SS) zone of Nigeria and the study area, in particular, might have worsened the land degradation and food production situation and led to widespread forced migrations. The literature on climate change and its resultant effects of desertification, drought, and famine especially in Nigeria, is replete with explanation and analyses of the causes and consequences of these problems almost to the exclusion of the social relations such as migration. Thus, although various studies have been carried out on droughts and their effects (Maryah 2004; Mortimore 1989; Nicholson and Jeeyoung 1997; Hulme 1992), the impact on forced migration has been minimally studied. This paper examines the impact of climate change on population mobility in two LGAs (Ngala and Kala-Balge) of N.E. Borno State, Nigeria.

## 3 Aim and Objectives

This paper aims at exploring how climate change and its resultant effects such as high-intensity droughts in the study area impact and operate to increase the tempo and frequency for forced out-migration.

The aim would be achieved through the following specific objectives:

1. To study the demographic, sociocultural, and economic background of ethnic groups who have migrated to other parts of Nigeria, Cameroon, and Chad Republics as a result of climate change.
2. Review the crucial linkages between climate change impact and forced migration.
3. To examine emigrants' contributions to the growth of their area of origin (the study area's) economy and development through remittances and investments.
4. Suggesting some appropriate strategies for drought adaptation and mitigation in the zone.

## 4 Justifications for the Study

This research is relevant and timely due to the problems emanating from climate change which is one of the current environmental hazards bedeviling the region. Looking at various efforts by governments at all levels over the years to alleviate poverty by providing employment to youths; it is an indication of understanding the devastating effects of the menace. Moreover, this study would serve as a guide and source of information for policy makers and researchers. A recent survey indicates that in sub-Saharan African countries, 31% of children aged between 5, 8, and above are engaged in various forms of labor such as slavery, trafficking, and forced recruitment for armed conflicts, militancy, migration to Europe by desert and the sea, and other hazardous work. Therefore, it becomes necessary to do research to avoid this kind of problem.

The most important aspect of this research is to ascertain to what extent the climate change affects population mobility in the study area. However, the study further aims at finding problems associated with climate change, its effects, and its implications on development as a result of forcing people out, so as to help the agencies, lawmakers, and other organizations that work with environmental problems in general. Another import of this research is that it will help to a great extent in understanding that climate change is one of the major push factors that repel people away from their original places of residence. It also serves as a stepping stone for further researches in the same field by broadening the scope and other related studies which can assist the policy makers (i.e., legislators) to formulate laws, which will at least in the long-term help to tackle the problem of climate change.

## 5 Scope

This study limits itself to a small part of the Sudano-Sahelian region of Nigeria. Moreover, concentrates on the effects of droughts along, and in only two Local Government Areas of Borno state, Nigeria. Migrants' destinations and how the resultant effects of climate change led to forced migration issue mainly into by this study.

## 6 Study Areas

The study area (Ngala and Kala-Balge LGAs) falls within the Sahelian region of North-Eastern Borno State of Nigeria (see Fig. 1). The Sahel region can be seen as the area of Africa lying between 12°N and 20°N. This area shares two climatic characteristics: one rainy season per year and August as the month of highest precipitation. Sahel is one co-climate zone located on the southern edge of the Sahara Desert. The Sahel covers part of the territory of (from west to east) Senegal, southern Mauritania, Mali, Burkina Faso, southern Algeria, Niger, Northern Nigeria, Chad Republic, etc. The whole length and breadth of the study area fall within the Sahel.

The name, Sahel, comes from the Arabic word for "border" or "margin". The region gained this name because it serves as the southern border to the Sahara. It is the transitional zone between the desert and the more tropical south central Africa.

The countries of the region are similar to climate, land cover, and economy. The Sahel is covered with brush, grasses, and stunted trees. The dominant industries are based on agriculture and herding livestock, much of which depends on rainfall. The Sahel region experiences strong year-to-year variations in climate. The rainfall varies from 200 to 600 mm a year (Akangbe et al. 2015). Over the past century, the annual precipitation has decreased. Most of the rain occurs during 3 to 4 months of summer.

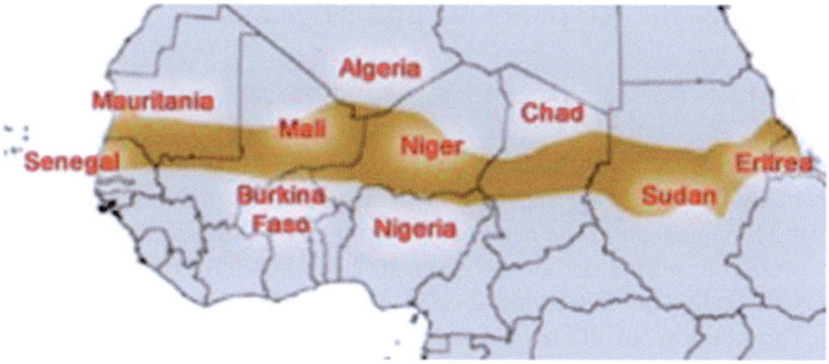


**Fig. 1** Map of West Africa showing Sahelian region

The study area has a total population of 297,332 comprising 153,387 males and 143,945 females in 2006 as follows: Ngala LGA has 122,127 males and 114,371 females and Kala-Balge has 31,260 males and 29,574 females.

## 7 Materials and Methods

Climate change is increasingly being recognized as a danger to food and human security through its potentialities of causing migrations. One of the potential trigger mechanisms is drought.

This study, in the field of climate change and its impact focuses on a socio-cultural aspect of forced migrants whose origin (homeland) is in Ngala and Kala-Balge LGAs of NE Borno State, Nigeria and have migrated to some parts of Cameroun Republic, Lake Chad, and Southern Nigeria, especially Lagos. Since most of the migrants were not physically present for the interactions, in most cases their family members were picked upon for the interview.

In each of the two LGAs, one urban and two rural settlements were selected, and 250 copies of the questionnaire were in all purposively administered. While 150 copies were administered in the urban areas (75 copies in each), 100 were administered in the rural areas (25 copies in each). A snowballing technique was used to identify the eligible persons for the survey (family members of emigrants). A questionnaire/interview schedule was administered to any adult in the house, male or female who volunteered to answer the questions. The data were analyzed using percentages which were viewed alongside conclusions drawn from reviewed.

The sample size of the present study consisted of 250 respondents purposively selected. A snowballing technique was in some cases used to identify the affected families. These were mainly family members of the migrants. The data was collected through "interview schedule" coupled with field observations and supplemented with secondary data.

## 8 Profiles of the Respondents and Migrants

This study shows that majority of the respondents (78%) were spouses or parents of the migrants. Thus, only about 22% of them are relatives, friends, wards, or in-laws of the migrants. Over two-thirds of them are having no or low education and are farmers or engaging in other unskilled labor work and other lower level white collar jobs. Sixty-nine percent of the respondents are living in nuclear families' houses. In terms of religious affiliation, all the 250 respondents in the study were found to be Muslims belonging to the ethnic groups of Kanuris, Shuwa-Arab, Kotoko, and others. The majority of the respondents (88%) were found in their youthful and working age group of 40–60 years. Most of them (62%) were capable or likely to become capable enough to purchase their own properties and live in their houses and about 65% of the respondents are married females (spouses) of the migrants. The remaining 40% are either parents or other relatives. They were made up of males (20%) and females (80%).

## 9 Results and Discussion

In appraising the emerging evidence of the manifestation of climate change in the North-East zone of Borno State, Nigeria, it has been ascertained that droughts are common occurrences. The land clearing done by CBDA for agricultural purposes has tremendously made the study area vulnerable to the effects of droughts and desertification. Temperature and rainfall variability have also been quite noticeable.

The sociocultural and economic background of the emigrants indicate that they are mainly Muslims, thought of different ethnic groupings they share similarities in culture and agriculture is their mainstay. They (15%) used to cultivate twice in a year and 20% is supplementing with fishing, trading, and herding.

In reviewing the crucial linkages between climate change impact and forced migration various evidence gotten from the field indicate that certainly most of the emigration (80%) took place as a result of that change.

Linkages/synergy between climate change and forced migration, pattern, and destinations of the migration are hereby discussed as per investigation carried out in the field:

The IASFM (2009: 1) designates that forced migration refers to "the movement of people displaced by natural or environmental disasters such as drought, land degradation, flooding, storms, heat waves, and so forth." Three categories of forced migrations have been identified; they are: conflict-induced, disaster-induced, and development-induced. What is of concern and is discussed in many kinds of literature is the last category. EACH-FOR (2008) clarifies that development-induced migrants are:

> People who are displaced from or who feel obliged to leave their usual place of residence because their lives, livelihoods, and welfare have been placed at serious risk as a result of adverse environmental, ecological or climate processes and events. The climate processes include: Climate Change, Global Warming. Land degradation, Rising Sea Levels, Deforestation, Soil erosion and Crop depletion. The climate events embrace earthquakes, volcanic eruptions, destructive storms, floods, tsunamis, droughts and famine EACH-FOR (2008: 8).

## 10 Linkages Between Drought and Forced Migration

When the two definitions are put together, it becomes fairly obvious that there is a linkage between Climate Change and forced migration; and between the latter and land degradation.

Among the positive impacts, 60% reported that regional differentiations which were found to be in country/place of origin are negligible or not being considered in host/destination country. Rather than regional, national identity is mostly noted. Most of those emigrating do, however, enjoy dual citizenship according to this study. These findings are also supported by a majority of the respondents. They said that those migrants who settle within the country did visit home more frequently than those in other countries even though they equally enjoy citizenship of those countries but the high transport fares. Again the majority of the respondents (80%) also reported receiving remittances from their heads of households (52%) and occasional financial help relatives or fellow family members in the time of need and crises (54%). Similarly, the majority of them reported about giving a substantial donation to the religious institutions, caste association for religious, educational, and social purpose and village for the development work in their village or home region (34%). Despite having these emotional bonds and socioeconomic ties, most of them (82%) reported their unwillingness to return home for resettlement due to their children's unwillingness for the same and physical discomfort resulting from the droughts in the study area. The findings are in agreement with Maryah's (2004) statement that climate change causes a lot of discomforts and land degradation.

## 11 Emigrants' Contributions to the Growth of Their Area of Origin

Emigrants' contributions to the growth of their area of origin (the study area's) economy and development through remittances and investments are tremendous.

The IASFM (2009) defines forced migration as "the movement of people displayed by natural or environmental disasters such as drought, land degradation, flooding, storms, heat waves, and so forth."

To examine emigrants' contributions to the growth of their area of origin (the study area's) economy and development through remittances and investments, the study shows that most of the male emigrants who left some members of their immediate families behind do make some remittances. These are evident in the way of life of the respondents. Some of them living around some water bodies do engage themselves in dry season agriculture using water pump machines purchased for them by their kin migrants.

## 12 Destinations of the Migrants

Migrants' destinations cut across Nigeria, Cameroun, and Chad Republics. They travel over 500 km in most cases. 40% of the migrants are said to have migrated to the Southern Nigeria, 55% of the fishing grounds of Cameroon such as Lac Doi,

Magwa, Mayo-Mbund, etc., and the remaining 5% to Lake Chad and Chadian part of the Lake Chad.

## 13 Causes of Migration in the Sahel

Causes of migration in the Sahel include declining in number of arable lands, diminishing territory and increasing in border disputes, environmentally induced migration as well as fragility and radicalization.

## 14 Conclusions and Recommendations

### *14.1 Conclusion*

This paper focuses on the climate change and population migration nexus. It is an exposition on the impact of climate change and its resultant consequences on forced migration. The paper has therefore brought to limelight the reasons given by the families of the migrants in the study area for the migration which are all attributable to droughts.

Having examined the impact of climate change on population mobility from N. E. Borno state to other parts of Nigeria or neighboring countries of Chad and Cameroon Republic, this study concludes by highlighting some of the measures that could help in coping with the vagaries of the changing climate.

### *14.2 Recommendations*

1. It is recommended that interbasin water transfers/dredging should be embarked upon from Chari river taking its source in Central Africa to river Ebeji/Surbewel. This would greatly improve the living standard of the people and reduce forced migration.
2. The use of shelterbelts and windbreaks is of great importance in this zone. This is because high wind velocities and soil erosionare common features of the zone which cause considerable damage to sown crops. Therefore, effects of high wind velocity and soil erosion could be minimized largely through the use of shelter plantation/afforestation.
3. Dry seed planting, information delivery, use of underground water should all be adopted.
4. Diversifying the economic base of the populace with emphasis on reducing overdependence on rain-fed agriculture, practicing multiple cropping, and changing site locally, for example, intensification of agriculture and use of fadama soils.

## References

Abaje IB, Ati OF, Iguisi EO (2012) Changing climate scenarios and strategies for drought adaptation and mitigation in the Sudano-Sahelian ecological zone of Nigeria. In: Iliya MA, Dankani IM (eds) Climate change and sustainable development in Nigeria, pp 99–121

Adejuwon JO (2006) Food crop production in Nigeria: potential effects of climate change. Climate Res 32:229–245

Adejuwon IS, Ogundele FO (2008): Climate change and health in Nigeria: incidence of vector-borne diseases (Malaria). In: Proceedings of 50th annual conference of the Association of Nigerian Geographers (ANG) held at the University of Calabar, 25–28 Aug 2008

Akangbe OM, Oluwabamidele M, Olalekan AB (2015) A critical analysis of the impact of climate change on migration: the Sahel Region in focus in national conference on migration, Kano Hall, Transcorp Hilton, Abuja

Akoteyon IS, Ogundele FO (2008) Climate change and Health in Nigeria: incidence of vector-borne diseases (malaria). In: Proceedings of 50th annual conference of the Association of Nigerian Geographers (ANG) held at the University of Calabar, 25–28 Aug 2008, pp 103–112

Aliyo SB (2008) An appraisal of climate change risks and institutional adaptations strategies in Kano State, Nigeria. In: Proceedings, 50th annual conference of the Association of Nigerian Geographers (ANG) held at the University of Calabar, 25–28 Aug, 2008, pp 103–112

Anyadike RNC (2009) Climate Change, national development and STI policy planning in Africa. Paper presented at the international conference and capacity building workshop on Africa' response to global challenges through science, technology and innovation, Abuja, Nigeria

Ayoade JO (1974) A statistical analysis of rainfall over Nigeria. J Trop Geogr 29:11–23

Ayoade IO (2004) Climate change. Vantage Publishers, Ibadan

Bashir D (2008) Sustainable management of drought and desertification in Nigeria. Report of the first national environmental summit on the theme: generating the environment for sustainable developing, pp 96–112. Federal Ministry of Environment. Housing and Urban Development

Bates BC, Kundzewicz ZW, Wu S, Alutikof JP (eds) (2008) Climate change and water. Technical paper of the Intergovernmental Panel on Climate Change, ICPC Secretariat, Geneva

EACH-FOR (2008) Environmental change and forced migrations scenarios policy-oriented research project of the European commission, 2007–2008, http://www.eachfor.eu/index.php, retrieved on 25 May 2009

Fisher G (2002) Climate change and agricultural vulnerability. IISA Publications

Gazali KAY (2005) The Kanuri in Diaspora: the contributions of Kanem-Borno Ulama to Islamic education in Nupe and Yoruba lands. CSS Press, 15, Broad Street, Lagos

Hulme M (1992) Rainfall changes in Africa 1931–1960 to 1961–1990. Int J Climatol 12:685–699

IASFM (International association for the Study of Forced Migration) (2009:1) Forced Migration online: what: file//c/Do.Migration.htm

IPCC (2001) (Intergovernmental Panel on Climate Change) Climate climate the IPCC scientific assessment. In: Houghton JT, Jenkins GJ, Ephramus JJ (eds) Cambridge University, Cambridge

IPCC (2007) Climate change, impacts, adaptation and vulnerability working group II contribution to the Intergovernmental Panel on Climate Change: summary for policy makers. IPCC Secretariat Geneva, Switzerland

Kowats RS, Campbell-Lendrum D, Matthies F (2005) Climate change and human health: estimating avoidable deaths and disease. Risk Anal 25(6):1409–1418

Mamman AB (2008) The challenge of change. Presidential address to the Association of Nigerian Geographer's 50th annual conference at the University of Calabar, Cross River State, Nigeria

Maryah UM (2004) Forage production and livestock management systems in the semi-arid zone of NE Nigeria. PhD seminar series, Bayero University, Kano

Maryah UM, Gadzama NM (1996) Temporal variability of rainfall and implications on agriculture and range management in the arid zone of N. E. Nigeria. In: Daura MM, Ibe BA, Ehiomere EO (eds) Monitoring the Nigerian environment

Mortimore M (1987) The lands of Northern Nigeria: some urgent issues. In: Mortimore M (ed) Perspectives on land administration and development in Northern Nigeria development of geography. Bayero University, Kano

Mortimore M (1989) Adaption to drought: farmers, famines and desertification in West Africa, Cambridge

Mshelia AD (2005) Adaptation strategies to climate change. J Energy Environ 17:74–90

Nicholson SE, Jeeyoung K (1997) The relationship of the El Nino/Southern oscillation to African. Int J Climatol 17(2):117–135

Odekunle TO, Andrew O, Aremu SO (2008) Towards a wetter Sudano-Sahelian ecological zone in twenty-first century Nigeria. Weather 63(3)

Oladipo EO (1993) Some aspects of the spatial characteristics of drought in Northern Nigeria. Natl Hazards, vol 8. Kluwer Academic Publishers, Netherlands, pp 171–188

Onoka'la PC (2011) Climate change and the need for curriculum development in geography. Niger Geogr J N Ser 7(2)

Solomon et al (2007) Climate change: the physical science basis. Contribution of working group to the fourth assessment report of the Intergovernmental Panel. Cambridge University Press, Cambridge, p 235

Waugh D (1995) Geography: an intergrated approach. Thomas Nelson and Sons, UK

# Part V
# Risk Assessment

# Detection of Areas Associated with Flash Floods and Erosion Caused by Rainfall Storm Using Topographic Attributes, Hydrologic Indices, and GIS

A. Bannari, A. Ghadeer, A. El-Battay, N.A. Hameed and M. Rouai

**Abstract** Storm floods are the most common of natural disasters that can affect infrastructure and cause human casualties, environmental destruction, and economical losses. Increased storm frequency and intensity related to climate change are aggravated by several factors, such as the growing occupation of floodplains, increased runoff from hard surfaces, inadequate management policy, and silted up drainage. The objective of this research is the integration of topographic attributes (elevation, slope, curvature, and water catchment), topographic profiles, and hydrologic indices derived from Digital elevation model (DEM) in a geographic information system (GIS) environment to detect areas associated with flash floods and erosion caused by rainfall storm and sediment transport and accumulation. The selected study area is the region of Guelmim city in Morocco. This region has been flooded several times over the past 30 years and was declared a "disaster area" in December 2014 after violent rainfall storms killed many people and caused significant damage to the infrastructure. GIS was used to extract topographic profiles and attributes as well as for the implementation of the stream power index (SPI), sediment transport index (STI), and compound topographic index (CTI). Moreover, it was used for spatial data management and manipulation whereas the PCI-Geomatica image processing system was used for fuzzy *k*-means unsupervised classification for topographic attributes and hydrologic indices. The obtained results show that hydrologic indices demonstrated that the rainfall and the topographic morphology are the major contributing factors for flash flooding and catastrophic inundation in the study area. The runoff water power delivers vulnerable topsoil and contributes strongly to the erosion and land degradation process, and then transports soil material and sediment to the plain areas through natural action, i.e. water power and gravity. The unsupervised classification leads to three homogeneous units of

A. Bannari (✉) · A. Ghadeer · A. El-Battay · N.A. Hameed
Geoinformatics Department, College of Graduate Studies,
Arabian Gulf University, Manama P.O. Box 26671 Kingdom of Bahrain
e-mail: abannari@agu.edu.bh

M. Rouai
Department of Earth Sciences, Faculty of Sciences,
University Moulay Ismail, BP. 11201, Zitoune, 50 000 Meknes, Morocco



S. Pirasteh and J. Li (eds.), *Global Changes and Natural Disaster Management: Geo-information Technologies*, DOI 10.1007/978-3-319-51844-2_13

dynamic response to hydrologic processes. The high levels of aggressiveness are encountered in the valleys and over areas with steeper slopes. The valleys are zones of flow accumulation receiving the contribution of large upslope drainage areas, thus allowing high rates of erosion. Conversely, low runoff aggressiveness is connected with areas of low slopes. Likewise, the role of the lithology associated with the terrain morphology is decisive in the erosion risk and land degradation in this region.

**Keywords** Soil erosion · Land degradation · Flood-storm · Runoff · Sediment transport · Climate change · GIS · Topographic attributes · Hydrologic indices · ASTER GDEM

## 1 Introduction

During the last four decades, the impact of climate change has become an undeniable reality, with a broad consensus of the international scientific community on the significance of its impacts on the environment, the economy, and society, especially in African countries (Cai et al. 2015). Morocco is a North African country with a dominant (mostly) semi-arid to arid climate and presents the typical characteristics of Mediterranean landscape vulnerable to land degradation processes, landslides, and desertification risks (Bannari et al. 2008; Maimouni et al. 2012). Currently, Morocco is experiencing the longest dry episode of its contemporary history characterized by a reduction of precipitation and a rise in temperatures. These weather conditions limit the growth of vegetation cover (sparse and scattered) especially in the High Atlas mountains. This situation in conjunction with human activities has led to degradation and the erosion of soil by wind and water. In addition, rainfall rarely occurs but with a high intensity during a short period of time, which causes flooding problems and accelerates the erosion phenomenon and land degradation. According to Erskine and Saynor (1996), catastrophic floods are defined as events with a flood peak discharge at a rate at least ten times greater than the mean annual flood. Heavy rains often induce floods in Morocco, including flash floods, rivers floods and mud floods during the rainy season. Indeed, in December 2014 violent storms caused flooding and impressive river floods in a large part of southern Morocco, especially in Guelmim city and regions located at the foot of the Anti-Atlas Mountains, which have peaks rising to over 2100 m. According to SIGMA (2015), this natural catastrophe caused the death of more than 46 persons and significant damage to the infrastructure villages were inundated causing thousands of houses to collapse, there was destruction of many oasis and agricultural fields (Fig. 1), power and telephone networks were damaged, and several bridges and roads were destroyed (Fig. 2). Total losses were estimated at approximately US$ 0.5 billion. Consequently, the region of Guelmim was declared a

**Fig. 1** Oasis, agricultural fields and a classroom inundation in Guelmim city region in December 2014 (Photos from the web)

**Fig. 2** Destruction of several roads and bridges' infrastructure in Guelmim city region in December 2014 (Photos from the web)

"disaster area" by the Moroccan government. This was not the first time that Guelmim city and the regions were devastated; these regions have been flooded several times over the past 30 years (January 1985, March 2002, and December 2014). Unfortunately, in addition to the climate change impact, flooding occurred because of the lack of emergency measures, a failure of development policy, the carelessness of citizens, as well as the Moroccan decision-makers have not learned from past lessons. However, it is very simple to understand and to know that the water always takes her bed!

Over the last century, climate change impacts on the environmental ecosystems coupled with the exponential growth of the human population have put significant pressure on soil resources (Cohen 1995). The climate change impacts contribute significantly in the risk of devastation from storm floods because climate is the principal driving force in hydrologic systems. At the global, regional and local scales, soil erosion by water, flood and runoff after a flood-storm are the most important contributing factors in the erosion and land degradation process (Eswaran et al. 2001) and, consequently, they form a real obstacle to sustainable development (Lal 2001; Vrieling 2007). Furthermore, to improve the management of the water regulation structures, there is a need to develop a methodology to maximize the water storage capacity and to reduce the risks caused by floods. Previous research activities have developed methodologies based on analyzing correlations between the agents of climate change, physiography, lithology, vegetation and land use (McDermid and Franklin 1994). Unfortunately, few of these parameters can be measured directly, but earlier studies have shown that information technology has gained increasing importance in providing useful substitute information. For instance, the extraction of topographic, hydrologic, and environmental information from the Earth observation satellite sensors is well developed (McDermid and Franklin 1994). Moreover, during the last two decades, geographic information system (GIS) and auxiliary data have become the fundamental solutions for flood monitoring and its impact assessment (Sanyal and Lu 2004). Likewise, topography information is considered crucial for flood hazard mapping, and a digital elevation model (DEM) is considered the most effective means to estimate flood depth from hydrologic indices or semi-empirical hydrological models (Pike et al. 2009).

Furthermore, the ability of rainfall to erode is a function of its intensity, duration, mass, and velocity of impact. This has been the concept behind the development of various *erosivity indices* to map regional and seasonal erosion risk (Morgan 2005). Indeed, the potential of topographic attributes and hydrologic indices were examined for their use in detecting areas affected by flood and erosion from heavy rainfall and inundation (Moore 1993). Scientists have proposed different topographic attributes (elevation, slope, orientation, and curvatures) and hydrologic indices, such as the catchment areas, flow accumulation, ridges, sediment transport and accumulation after flooding or stream channels, compound topographic index (CTI), stream power index (SPI), and sediment transport index (STI) (Moore 1993; Dymond and Harmsworth 1994). Geology, geomorphology, topography, tectonics, soil, vegetation cover, precipitation, climate, river incision, channel stability, and landscape evolution have been involved in the conceptualization of these indices (Florinsky 2012; Bagnold 2015). The objective of this research is the integration of topographic profiles and attributes, and hydrologic indices derived from DEM in a GIS environment to characterize the areas associated with flash floods and erosion caused by rainfall storm, and spatial sediment transport and accumulation distribution.

## 2 Materials and Methods

The used methodology is summarized in Fig. 3. It involves the fundamental steps of DEM preprocessing, then topographic profiles and attributes, and hydrologic indices retrieval. Subsequently, in the PCI-Geomatica image processing system (PCI-Geomatics 2015), fuzzy *k*-means unsupervised classification of these variables was considered for the landscape segmentation into different units of soil erosion and sediment accumulation.

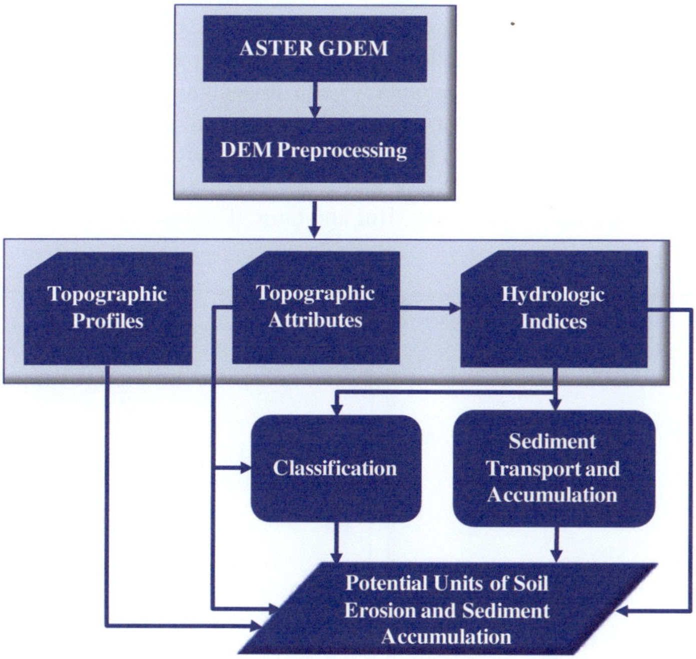


**Fig. 3** Methodology flowchart

## 2.1 Study Site

Guelmim is a city in south of Morocco, often called *Gateway to the Desert*; it is the capital of the Souss-Massa-Drâa region (28° 59′ 02″N, 10° 03′ 37″W). It is located at the foot of the western Anti-Atlas mountains with peaks rising to over 2100 m and it follows the course of underground shallow aquifers and dry rivers (Fig. 4). Settled agriculture developed around underground water sources and dry rivers beds that flood during the rainy season. It is characterized by a semi-arid and arid subtropical climate. Temperature range varies from 12 °C in January to 49 °C in July. Annual rainfall averages between 70 and 120 mm/year (Dijon and El Hebil 1977). The Assaka River drains a large watershed on the southwest border of the Anti-Atlas and its Saharan edges in the Guelmim area (Weisrock et al. 2006). Its watershed covers approximately 7000 km$^2$ and develops on the southern slopes of the Anti-Atlas (Wengler et al. 2002). In the lower part of its course, the river valley shows the stepping and the nesting of several alluvial terraces intersecting the Appalachian relief of the Anti-Atlas mountains and depressions which offer a landscape of hills and small mountains depending on the resistance of rocks to erosion. The Assaka River, the confluence of three wadis: Seyyad, Noun and Oum Al-Achar, crosses the last folded chains of the Anti-Atlas before flowing into the Atlantic Ocean. In this part of its course, it has aggraded a large system of alluvial terraces. The geological formations that feed alluvium are granite, schist, quartzite, sandstone, limestone, dolomite, marl, conglomerate, andesite, and rhyolite (Fig. 5). They correspond to the top of the Precambrian (Ifni boutonniere) and lower Paleozoic rock, which forms part of the Anti-Atlas sedimentary cover (Wengler et al. 2002). From a geological point of view (Choubert 1963), the study region constitutes a complex synclinal, framed and surrounded in the N, W and S by three Precambrian anticlinal inlets (boutonnieres): Kerdous-Tazeroualt, Ifni and Guir. The two mean structural unities

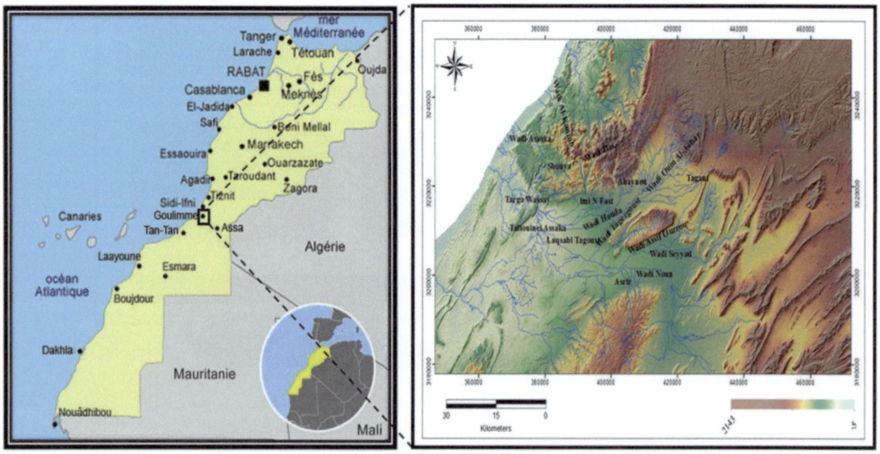


**Fig. 4** Map of the study site location (Guelmim city region, Morocco)

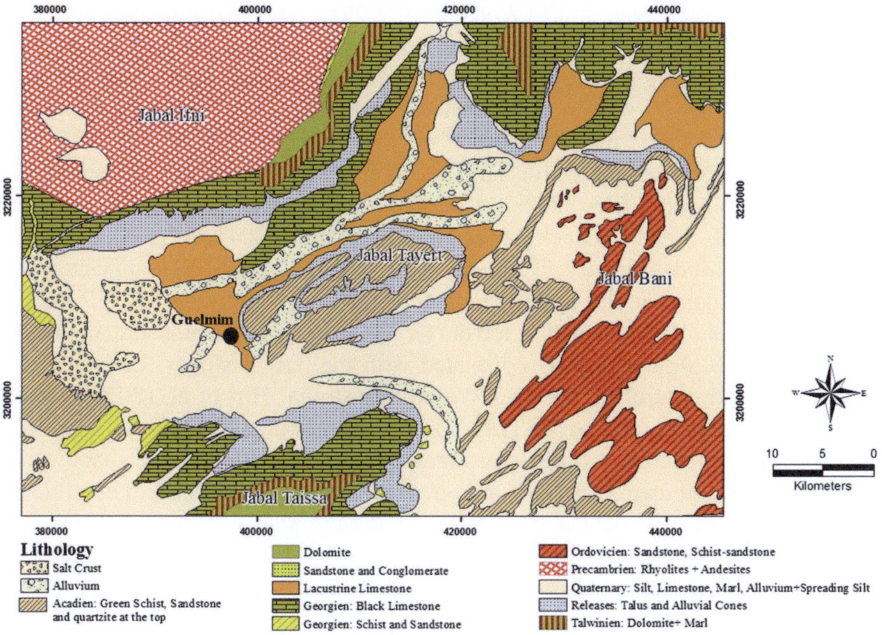


**Fig. 5** Lithologic map of the study site

in the region are the carbonate plateaus and the folded Bani hills. The most important Infra-Cambrian and Cambrian carbonate plateaus are located north, consisting of a continuous area bordering from W to E the Ifni Inlet, Akhsass plateau and the southern flank of the Kerdous inlet. The second one, located south, is formed by the external part of Jabal Guir-Taissa. These plateaus are surrounded by schist and sandstone formations of the Georgian age. At their foot begin large and elongated plains, named 'feijas' and consisting of Acadian schist covered by Quaternary deposits (mainly lacustrine carbonate and silts). At the centre of the Guelmim basin lies Jabal Tayert which is formed by green Upper Acadian schist and which is covered at the top by hard sandstone and quartzite bars. The Bani Jabal is a folded structure consisting of several aligned and NE-SW oriented synclinals alternating with narrow anticlinals formed by Acadian or Ordovician sandstones and quartzites.

## 2.2 ASTER GDEM Data

The ASTER (Advanced Spaceborne Thermal Emission and Reflection Radiometer) GDEM (Global Digital Elevation Model) is a joint product developed and made available to the public by the Ministry of Economy, Trade, and Industry (METI) of Japan and the United States National Aeronautics and Space Administration (NASA). It is generated from data collected from the optical instrument ASTER onboard the TERRA spacecraft (Welch et al. 1998). This instrument was built in December 1999 with an along-track stereoscopic capability using its nadir-viewing and backward-viewing telescopes to acquire stereo image data with a base-to-height ratio of 0.6 (NASA 2014). Since 2001, these stereo pairs have been used to produce single-scene (60 × 60 km) DEM based on the stereo-correlation matching technique using UTM map projection and WGS84 geodetic reference (Hiranoa et al. 2003). According to Chrysoulakis et al. (2004), the accuracies of the produced ASTER GDEM over the Greek islands are ±15 and ±12.41 m, respectively, for the planimetry and altimetry. They considered these accuracies satisfactory for watershed management and hydrological applications. Testing the ASTER GDEM over Vancouver (West Canadian territory), Toutin (2002) demonstrated that the derived DEM is almost linearly correlated with the terrain slopes. Using a set of geodetic ground control points over Western Australia, Hirt et al. (2010) showed that the vertical accuracy of ASTER DEM is approximately ±15 m. They also reported that this accuracy varies as a function of the terrain type and shape, and it is relatively low in areas with low topographic variability. In 2011, the validation and the accuracy assessment of the ASTER GDEM products (version-2) were made jointly by U.S. and Japanese partners (ERSDAC 2011; NASA 2011). The results of this study showed that the absolute geometrical calibration accuracies expressed as a linear error at the 95% confidence level are ±8.68 and ±17.01 m, respectively, for planimetry and altimetry (Meyer 2011). Consequently, the DEM over the Guelmim region (Fig. 6) was downloaded from USGS data explorer gate (USGS

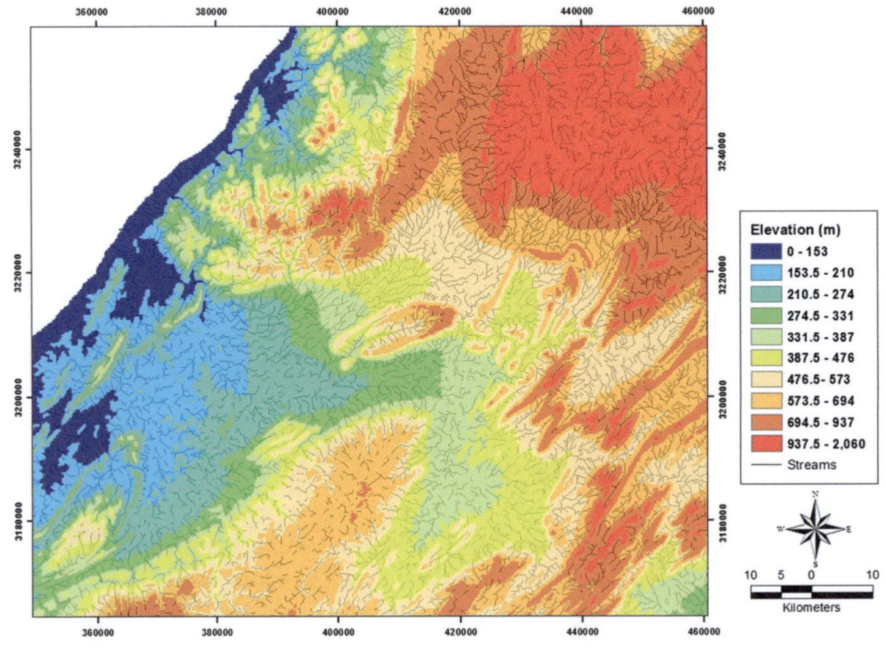


**Fig. 6** ASTER DEM over the study site

2015) and was preprocessed using ArcGIS (ESRI 2015); then it was used in this study for topographic profiles and attributes, and hydrologic indices derivation.

## 2.3 Sediment Transport and Hydrologic Indices

Topography represented by a DEM can be used to model flood impact on land erosion and degradation, and sediment accumulation because water tends to flow and accumulate in response to gradients in gravitational potential energy (Murphy et al. 2009). Moreover, it has a vital role in the spatial variation of hydrological conditions such as soil moisture, groundwater flow, and slope stability. Topographic attributes have, therefore, been used to describe spatial soil elevation, slope and orientation (Moore et al. 1991). Scientists have proposed many hydrologic indices to map soil erosion, soil sediments, and soil moisture conditions based on continuous field models (Grunwald 2006), in which the various properties are mapped considering certain numbers of pixels with continuous coverage across the landscape. This approach has been greatly supported by the development of computer science, GIS technologies, and image processing systems (Moore et al. 1991; Moore 1996; Murphy et al. 2009). According to Kothyari et al. (2002), there are many relationships available in the literature to estimate sediment transport. They reported that the best suited for integration with GIS is the relationship proposed by Moore and Wilson (1992), based on unit stream power theory and called the STI. It is a non-linear function of specific discharge and slope and it is derived by considering the transport capacity limiting sediment flux and catchment evolution erosion theories (Moore and Wilson 1992). This index is a fundamental factor in the Universal Soil Loss Equation (USLE) (Wischmeier and Smith 1978). It calculates a spatially distributed sediment transport capacity and may be better suited to landscape assessments of erosion than the original empirical equation because it explicitly accounts for flow convergence and divergence (Moore and Wilson 1992; Desmet and Govers 1996). It has been well tested under different DEM resolutions (Zhang and Montgomery 1994), and different flow routing methods (Quinn et al. 1991). Another semi-empirical index within the runoff model is the soil wetness index (SWI), named also the topographic wetness index (TWI) or the CTI, originally developed by Beven and Kirkby (1979). It was used extensively to quantify the topography effects on hydrological processes. It indicates quantitatively the balance between water accumulation and drainage conditions at the local scale. Otherwise, it describes the tendency for a site to be saturated at the surface given its contributing area and local slope characteristics. Furthermore, the SPI, which is a semi-empirical model, was developed and used to describe the ability to transfer sediment in channel streams, to estimate the sediment rate in basin hydrology, and to assess the flood risks. These hydrological indices were used for mapping landslide susceptibility in the Himalayas of Nepal (Poudyal et al. 2010). They are considered in this research, and their equations are as follows:

$$\mathrm{STI} = [(m+1).(A_s/22.13)^m.(\sin\beta/0.0896)^n] \quad (1)$$

$$\mathrm{CTI} = [\ln((A_s+0.001)/((\beta/100)+0.001)] \quad (2)$$

$$\mathrm{SPI} = [\ln((A_s+0.001).((\beta/100)+0.001)] \quad (3)$$

Where, "$A_s$" is the flow accumulation, "$\beta$" is the slope, "ln" is the Napierian logarithm, $m$ = 0.4 and $n$ = 1.3. "$A_s$" determines how much water is accumulating from upstream areas and therefore identifies areas that contribute to overland flow. According to several tests, a value of 200 m attributed for "$A_s$" have provided optimal results. In order to characterize the sediment transport and accumulation, and to evaluate the soil erosion and degradation areas affected by floods resulting from a heavy rainfall and inundation, we considered an integration analysis approach based on the topographic profiles and attributes, and hydrologic indices in a GIS environment. Then, according to Haboudane et al. (2002), a simplified approach to characterize erosion potential and runoff power through fuzzy *k*-means unsupervised classification of topographic attributes, and hydrologic indices was adopted.

## 3 Results Analysis

It is greatly pertinent to make a clarification on the hydrographic network and drainage system of the study area because a significant hydrological network is essential for flow direction modeling and spatial analysis. The Guelmim watershed covers a total area of approximately 7000 km$^2$, forming a network of wadis (rivers), along which there are several spreading floodwater areas (Fig. 4). The hydrographic network is made up of 3 sub-watersheds of the following main wadis: Wadi Seyyad, Wadi Noun, and Wadi Oum Al-Achar (Wengler et al. 2002; Naser 2006). Wadi Seyyad originates at 1200 m on the southern slopes of the Anti-Atlas mountains. It flows in an E-W direction, and mainly receives numerous tributaries of its right bank; its watershed is about 2860 km$^2$. Wadi Noun drains the southern area, where the bit marked with river beds which promote natural flooding. With a length of 143 km, its watershed comprises about 2240 km$^2$. Finally, Wadi Oum Al-Achar, with a watershed of 1170 km$^2$, crosses a wide plain of 7 km and is located between the Tayert hill and Ifni boutonnière. It drains the southern slopes of the Akhsass region, and its main tributaries are located in the plains. All three wadis lie on schistous impermeable large valleys or feijas, covered by low permeable Quaternary carbonates and fluvio-lacustrine silts. The confluence of the three wadis, downstream from Guelmim city, forms Wadi Assaka, which begins in the Akhsass massif and reaches 1150 m altitude. It goes through the corridor between the Jabal Adrar and then along Guelmim west, eventually discharging into the Atlantic Ocean after crossing narrow gorges. This hydrographic system is often inactive especially during the summer, when the flow is very low, but it becomes active during the winter period (December to March). This configuration is the cause of many talwegs and wadis draining the area. All the runoffs are thus directed automatically to the city of Guelmim which is subject to a hydrological regime unregulated surface.

The study site has two main geomorphologic units, the limestone plateaus of the Anti-Atlas and the Bani quartzite ridges, as natural barriers (Figs. 4 and 5). It is characterized by broader valleys and depressions (feijas) surrounded by hills with height varying from 153 to 2060 m (Fig. 6), and strong slopes varying between 9.5° and 26° which are converging towards the interior of the Guelmim plain. The topography of this plain is classified into seven classes, whose altitude range vary significantly from 200 to 573.5 m starting from northeast to southwest, with approximately 373.5 m height difference. This morphology leads to water retention in the case of rainstorm and, consequently, contributes to the risk of inundation. Figure 7 illustrates three topographic profiles; two from the NE to SW and one from NW to SE, showing this natural barrier, which leads to water retention in the case of high precipitation intensity. Thus, it is one of the factors supporting the risk of flash floods. The profile "A" from NW to SE shows an appropriate morphology for runoff water catchment with a depth of 250 m between two mountains which surround the plains in that direction (Jebel Ifni and Jebel Taissa). Figure 7b illustrates the topographic profile "B" from the NE to SW direction showing the area prone to inundation. Indeed, we see that the slope orientation and direction of the Guelmim watershed are facing the center of the plain. Also, we observe a hill in the East that forms a natural barrier with a denivelation of approximately 100 m creating a natural basin promoting the accumulation of water and sediments over 8 km distance. The topographic profile "C" (Fig. 7) starting at the foot of Guelmim city illustrates a very strong slope (26°), which ends on a terrain with concave morphology forming a natural basin. This basin facilitates the accumulation of storm floods, thereby concentrating runoff water, sediment, and mud load.

Furthermore, the hydrologic indices based on the topographic attributes (slope, aspect, and flow accumulation) and linked to the drainage system analysis demonstrate the potential of STI, CTI, and SPI to model flood impact on land erosion and degradation, and sediment accumulation. Figure 8 shows the STI-derived map with values varying between 0 and 298. It illustrates the spatial distribution of the sediment transport capacity and accumulation. The highest values (150–298) are related to the steep slopes and ridges corresponding to "feija" schist and soft Quaternary deposits, ravines, and water streams. These areas are associated with a significant degree of sediment transportation and, consequently, significant soil erosion and degradation. This index shows the increased erosivity of channel flow in the downstream, and sediment deposition in the plain. The values nearest to zero are associated with very hard like Precambrian magmatic rocks (andesite and rhyolite), Adoudounian limestone and dolomite, Acadian sandstone and Ordovician quartzite, which are located at the top of the mountains (Fig. 5). These areas are characterized with different rock types and they are subject to different degrees of erodibility, but in general it is associated with a low erosion risk. However, the values ranging from 5 to 25 are located in the plain progressively

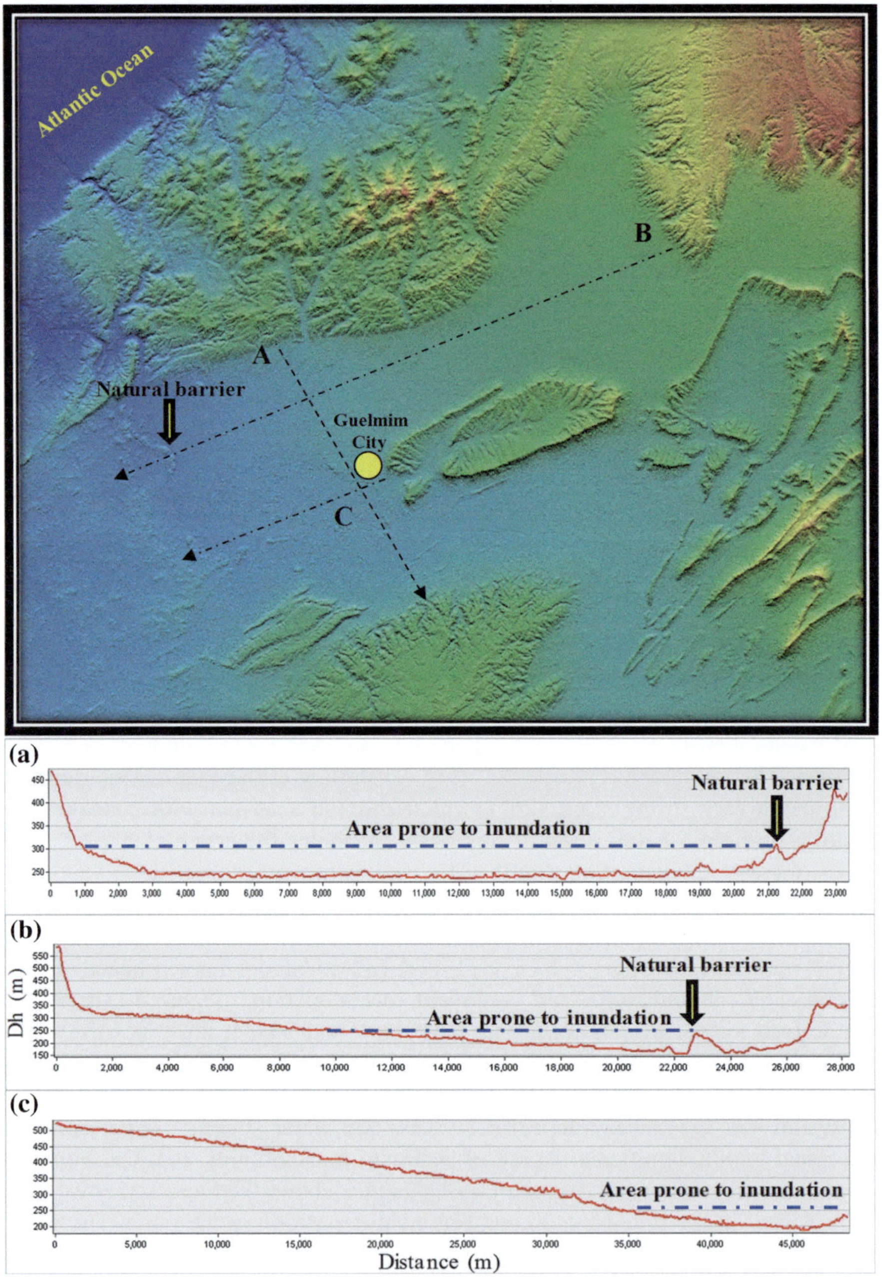


**Fig. 7** Three topographic profiles, from the NW to SE (**a**) and from the NE to SW (**b** and **c**) illustrating the natural barrier, which leads to water retention

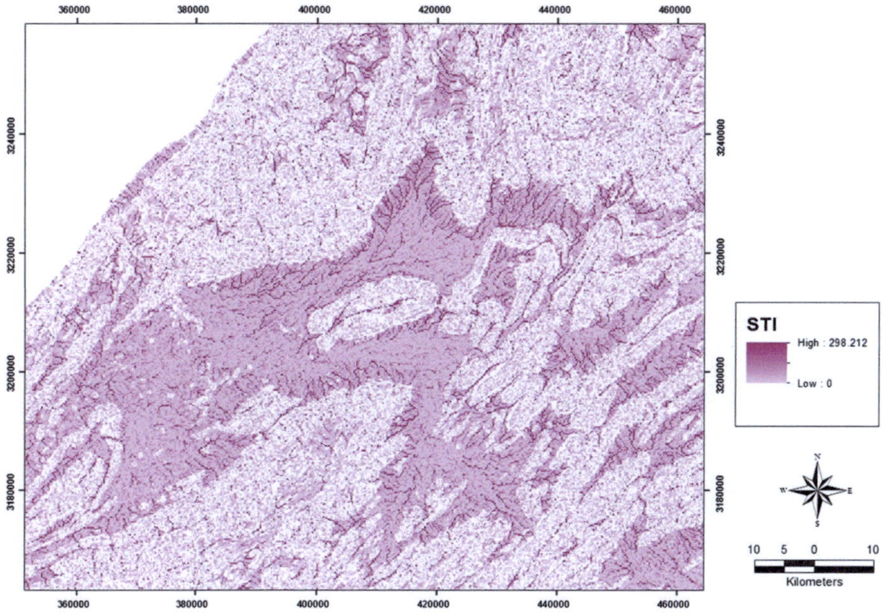


**Fig. 8** Sediment transport index and accumulation map

following the slope and the flow accumulation from the NE to the SW. At the very low slopes in the plain, the STI values around 3.5 reflect the slow mobility and, consequently, the accumulation of the sediment. This index is well correlated with landscape erosion assessments because it explicitly demonstrates the sediment flow convergence and divergence from the top of the mountains to the areas prone to inundation and sediment accumulation.

Figure 9 illustrates the CTI derived map. These index values range between −5.9 and 15.6, and it highlights the hydrological system unlike STI. We see that the higher an area's values, the greater the potential for that area to be saturated with water based upon its contributing neighborhood and local slope characteristic, which is low. Furthermore, high CTI values reveal the potential of pixels to be wet before other surrounding and contributing pixels. Therefore, areas with high CTI values are more susceptible to flash flooding and inundation as compared to those with low CTI. The areas with negative and very low CTI values are characterized by high altitudes and strong slopes highlighting the local drainage system conditions. The intermediate values are related to the steep slopes and ridges, ravines, and water streams, which are the most important contributing factors to erosion and the land degradation process (talus and alluvial cones, Quaternary lacustrine limestone, spreading silts, and salt crusts; see Fig. 5). Globally, this index shows clearly how the water flows down the slopes over a long distance, then the vast morphology of the watershed and the opposite side slopes support the floods accumulation, runoff,

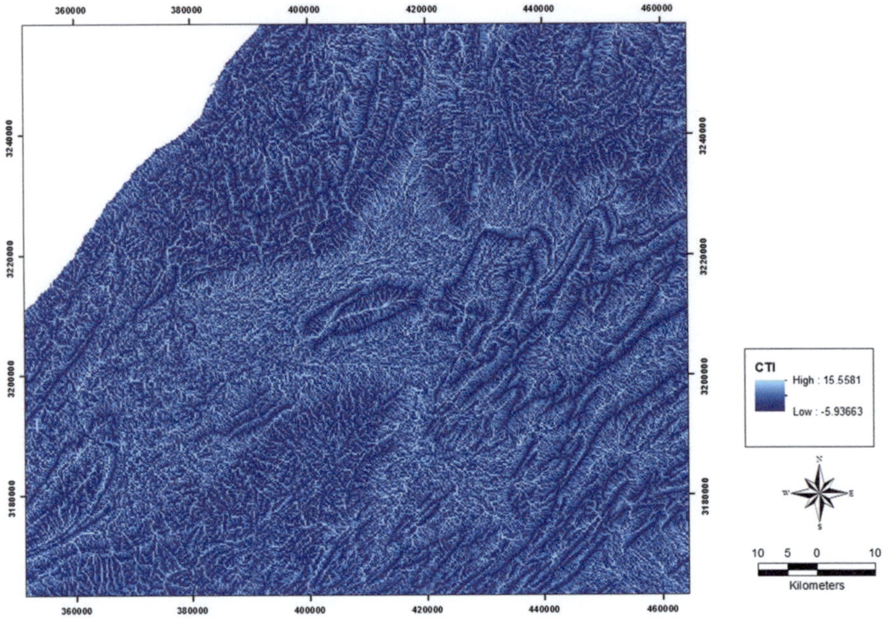


**Fig. 9** Compound topographic index map

and sediment load. Undeniably, the lowest and flattest areas were affected most by the flash floods because water tends to flow and accumulate in response to gradients in gravitational potential energy.

The SPI describes the ability to transfer sediment in a channel's streams and evaluate the flood risks in basin hydrology similarly to the CTI (Fig. 10). It shows negative values for areas with topographic potential for deposition and positive values for potential erosive areas. This index map shows areas with the highest values (approximately 1–5.6) representing the streams and the drainage system, depressions, and broader-valleys situated at height levels of between 500 and 2100 m, and related to a strong slope gradient ($\geq 20°$). This terrain morphology contributes significantly to the erosion's aggressiveness and land degradation risk process. The SPI values around −9 are located at the top of the mountain with a significant slope gradient varying from 9° to 12°. These zones are characterized by hard rock's such as Precambrian quartzite, Adoudounian limestone and dolomite, Ordovician quartzite and sandstone, and Georgian black limestone. The lowest values (between −13.8 and −2) represent relatively flat areas with a low slope ($\leq 4°$) from the NE to the SW direction, which is the hydrographic network direction, and morphological factors influencing the susceptibility to flooding and sediment deposition and accumulation. We can observe that the SPI and the CTI reflect the tendency of water accumulation in the landscape and highlight areas prone to both fast moving and pooling water.

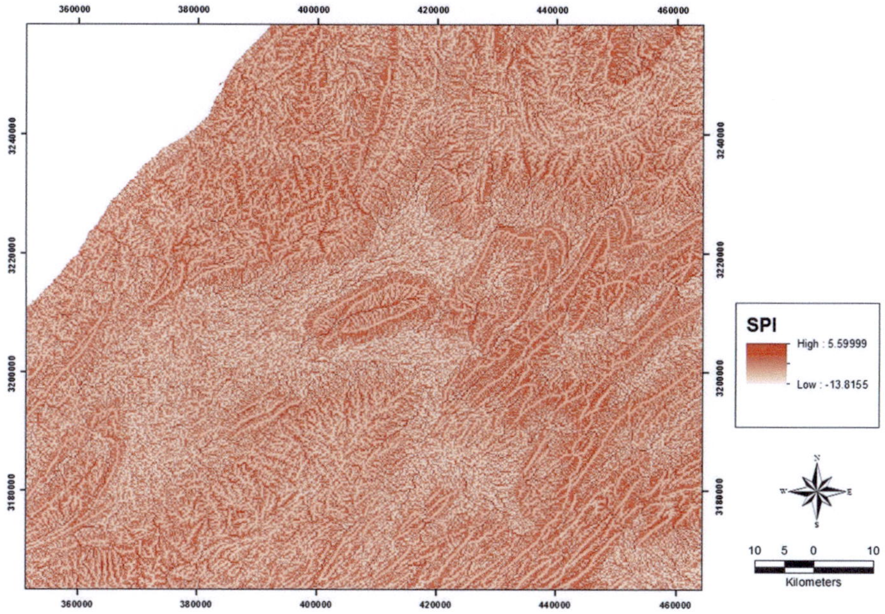


**Fig. 10** Stream power index map

In general, these three hydrologic indices demonstrate that rainfall and topographic attributes are the major contributing factors to flash flooding and catastrophic inundation. These conclusions are in agreement with Marchi et al. (2010) results, in which they characterized the extreme flash floods and implications for flood risk management in Europe. According to these indices derived maps, after a flood-storm in the Guelmim basin, the runoff water power delivers vulnerable topsoil and contributes strongly to the erosion and land degradation process, and then transports soil material and sediment to the plain through the natural action, i.e., water power and gravity. As illustrated by the photos (Figs. 1 and 2) which were acquired during the same day of the flood storm, the water color was dark red because of its turbidity as it was very rich with delivered sediments and eroded particles. In addition, the classroom of a primary school was filled with mud after the water retreated. Certainly, the role of the lithology associated with the terrain morphology is decisive in the erosion risk and land degradation in this region.

An automatic classification of these topographic attributes and hydrologic indices leads to six classes, which were regrouped into three homogeneous units of dynamic response to hydrologic processes. The regrouping has been performed using the fuzzy *k*-means unsupervised classification. These classes were qualified in terms of surface flow aggressiveness, as shown in Fig. 11, which illustrates the map of areas with similar potential for erosion and sediment accumulation units over the

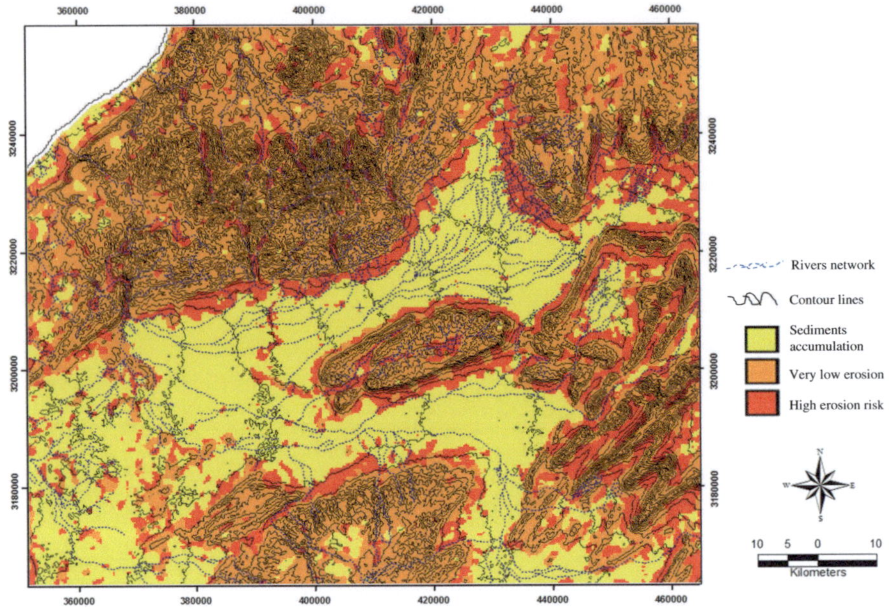


**Fig. 11** Map of erosion potential and sediment accumulation units obtained by classification of topographic attributes and hydrologic indices, with superposition of topographic contours and hydrographic network

watersheds of the Guelmim region, with superposition of topographic contour lines and hydrographic network. In this Figure, high levels of aggressiveness are encountered in the valleys and over areas with steeper slopes, as underlined by topographic contours. The valleys are zones of flow accumulation receiving the contribution of large upslope drainage areas, thereby allowing high rates of erosion. Conversely, low runoff aggressiveness is connected with areas of low slopes and small drainage areas located in Piedmont and in the higher mountainous zones. Moreover, by reference to the lithological map (Fig. 5), the areas classified with a high potential risk of erosion (Fig. 11) are formed with Acadian schist characterised by high erodibility, Adoudounian fissured carbonate cliffs in the north, Georgian black limestone on the Taissa Jabal border, and Ordovician Bani sandstones with very low vegetation recovery or completely absent. The areas with very low erosion are characterized by volcanic Precambrian rocks (rhyolite and andesite) which are resistant to erosion and also which form the highest peak of the mountains (Ifni boutonniere), Ordovician Bani quartzite crests and some synclinal throughs (Tayert).

# 4 Conclusions

At the global, regional, and local scales soil erosion by water, flood, and runoff after a flood-storm are the most important contributing factors in the erosion and land degradation process, especially in developing countries. During the last century, climate change impacts on the environmental ecosystems coupled with the exponential growth of the human population have put enormous pressure on soil resources and have threatened sustainable development. The objective of this research was the integration of topographic profiles and attributes, and hydrologic indices derived from DEM in a GIS environment to detect areas associated with flash floods and erosion caused by rainfall storm, sediment transport and accumulation in the region of Guelmim city in Morocco. In addition, the PCI-Geomatica image processing system was used for fuzzy *k*-means unsupervised classification for topographic attributes and hydrologic indices (STI, CTI and SPI). The obtained results show that hydrologic indices demonstrated that the rainfall and the topographic morphology are the major contributing factors in flash flooding and catastrophic inundation in the study area. The runoff water power delivers vulnerable topsoil and contributes strongly to the erosion and land degradation process, and then transports soil material and sediment to the plain areas through natural action. The unsupervised classification leads to the homogeneous units of dynamic response to hydrologic processes. The high levels of aggressiveness are encountered in the valleys and over areas with steeper slopes. The valleys are zones of flow accumulation receiving the contribution of large upslope drainage areas, thus allowing high rates of erosion. On the other hand, low runoff aggressiveness is connected with areas of low slopes. Likewise, the role of the lithology associated with the terrain morphology is decisive in the erosion risk and land degradation in this region. Certainly, the synergy of this derived information from DEM with multi-temporal remote sensing data allows the robustness of the erosion process modeling and analysis. In addition, the application of remote sensing and GIS is clearly a very efficient and cost effective way of flood management in developing countries.

**Acknowledgements** The authors would like to thank the Arabian Gulf University (AGU) for their financial support. We would like to thank the LP-DAAC NASA-USGS for the ASTER GDEM data. Our gratitude to many people who have made the used photos available on the web for consultation and public use. Finally, we express gratitude to the anonymous reviewers for their constructive comments.

# References

Bagnold RA (2015) An approach to the sediment transport problem from general physics. Physiographic and hydraulic studies of rivers, United States Geological Survey Professional, Paper-422I. Available online: http://www.uvm.edu/~wbowden/Teaching/Stream_Geomorph_Assess/Resources/Private/Documents/1966_Bagnold_river_sediments.pdf. Accessed Sept 2015

Bannari A, El-Harti A, Haboudane D, Bachaoui M, El-Ghmari (2008) A Intégration des variables spectrales et géomorphométriques dans un SIG pour la cartographie des zones exposées à l'érosion. Revue Télédétection, vol 7(1-2-3-4), pp 393–404

Beven KJ, Kirkby MJ (1979) A physically-based variable contributing area model of basin hydrology. Hydrol Sci Bull 24:43–69

Cai X, Zhang X, Noël PH, Shafiee-Jood M (2015) Impacts of climate change on agricultural water management: a review. WIREs Water. doi:10.1002/wat2.1089

Choubert G (1963) Histoire géologique du Précambrian de l'Anti-Atlas. Tome1. Notes et Mémores, Serv. Géol. Maroc, N° 162, 352p

Chrysoulakis N, Abrams M, Feidas H, Velianitis D (2004) Analysis of ASTER multispectral stereo imagery to produce DEM and land cover databases for Greek Islands: the real DEMs project. In: Prastacos P, Cortes U, De-Leon JL, Murillo M (eds) Proceedings of environment progress and challenges, pp 411–424

Cohen JE (1995) Population growth and Earth's human carrying capacity. Science 269 (5222):341–346

Desmet PJJ, Govers G (1996) Comparison of routing algorithms for digital elevation models and their implications for predicting ephemeral gullies. Int J Geogr Inf Sci 10:311–332

Dijon R, El Hebil A (1977) Bassin de l'oued Noun et basins côtiers d'Ifni au Draa. In: Ressources en eaux du Maroc, Notes et Mémoires du Serv. Géol. Maroc, N° 231, Tome3, 438p

Dymond JR, Harmsworth GR (1994) Towards automated land resource mapping using digital models. ITC J 2:129–138

Erskine WD. Saynor MJ (1996) Effects of catastrophic floods on sediment yields in southeastern Australia. Erosion and sediment yield: global and regional perspectives. In: Proceedings of the Exeter symposium, IAHS Publication, vol 236, pp 381–388

ERSDAC (2011) ASTER-GDEM version 2, validation report, Japan's validation report, 24. Available online: https://www.jspacesystems.or.jp/ersdac/GDEM/ver2Validation/Appendix_A_ERSDAC_GDEM2_validation_report.pdf. Accessed 15 March 2015

ESRI (2015) Getting to know ArcGIS. In: Law M, Collins A (eds) , 4th edn, 808 pages. Available online: http://esripress.esri.com/bookresources/index.cfm?event=catalog.book&id=16. Accessed 15 Feb 2015

Eswaran H, Lal R, Reich PF (2001) Land degradation: an overview. In: Bridges EM, Hannam ID, Oldeman LR, Penning-de-Vries FWT, Scherr SJ, Sombatpanit S (eds) Response to land degradation. Science Publishers Inc., Enfield, NH, USA, pp 20–35

Florinsky IV (2012) Digital terrain analysis in soil science and geology. Academic Press is an imprint of Elsevier. The Boulevard, Langford Lane, Kidlington, Oxford OX5 1GB, UK, 379

Grunwald S (2006) What do we really know about the space-time continuum of soil- landscapes? In: Grunwald S (ed) Environmental soil-landscape modeling. CRC Press, Taylor & Francis Group, Boca Raton, pp 3–36

Haboudane D, Bonn F, Royer A, Sommer S, Mehl W (2002) Land degradation and erosion risk mapping by fusion of spectrally based information and digital geomorphometric attributes. Int J Remote Sens 23(18):3795–3820

Hiranoa A, Welcha R, Langb H (2003) Mapping from ASTER stereo image data: DEM validation and accuracy assessment. ISPRS J Photogramm Remote Sens 57:356–370

Hirt C, Filmer M, Featherstone WE (2010) Comparison and validation of the recent freely available ASTER-GDEM ver1, SRTM ver4. 1 and GEODATA DEM-9S ver3 digital elevation models over Australia. Aust J Earth Sci 57(3):337–347

Kothyari UC, Jain MK, Ranga-Raju KG (2002) Estimation of temporal variation of sediment yield using GIS. Hydrotogical Sci J des Sci Hydrol 47(5):693–706

Lal R (2001) Soil degradation by erosion. Land Degrad Dev 12(6):519–539

Maimouni S, Bannari A, El-Hrati A, El-Ghmari A (2012) Potentiels et limites des indices spectraux pour caractériser la dégradation des sols en milieu semi-aride. J Can de Télédétec 37 (3):285–301

Marchi L, Borga M, Preciso E, Gaume E (2010) Characterization of selected extreme flash floods in Europe and implications for flood risk management. J Hydrol 394(1–2):118–133

McDermid GJ, Franklin SE (1994) Spectral, spatial and geomorphometric variables in remote sensing of slope processes. Remote Sens Environ 49(1):57–71
Meyer D (2011) ASTER GDEM version 2—Summary of validation results, 27, Available online: https://lpdaacaster.cr.usgs.gov/GDEM/Summary_GDEM2_validation_report_final.pdf. Accessed 25 June 2015
Moore ID (1996) Hydrologic modeling and GIS. In: Goodchild MF, Steyaert LT, Parks BO, Johnston C, Maidment D, Crane M, Glendinning S (eds) GIS and environmental modeling: progress and research issues. GIS World Books, Fort Collins CO, pp 143–148
Moore ID, Wilson JP (1992) Length-slope factors for the Revised Universal Soil Loss Equation: simplified method of estimation. J Soil Water Conserv 47:423–428
Moore ID, Grayson RB, Ladson AR (1991) Digital terrain modelling: a review of hydrological, geomorphological, and biological applications. Hydrol Process 5:3–30
Moore ID, Gessler PE, Nielsen GA, Peterson GA (1993) Soil attribute prediction using terrain analysis. Soil Sci Soc Am J 57:443–452
Morgan RPC (2005) Soil erosion and conservation, 3rd edn. Blackwell Science Ltd a Blackwell Publishing company, p 304
Murphy PNC, Ogilvie J, Arp P (2009) Topographic modelling of soil moisture conditions: a comparison and verification of two models. Eur J Soil Sci 60:94–109
NASA (2011) ASTER Global Digital Elevation Map Announcement. Available online: https://asterweb.jpl.nasa.gov/gdem.asp. Accessed 15 March 2015
NASA (2014) METI and NASA Release the ASTER Global DEM. Available online: https://lpdaac.usgs.gov/about/news_archive/meti_and_nasa_release_aster_global_dem. Accessed 25 June 2014
Naser N (2006) Utilisation des Systeme d'Information Géographiques en Hydrogéologie en Vue de l'éboration d'un Outil de Gestion des Ressources en Eau du Bassin de Guelmim. Institut Agronomique et Veterinaire - Hassan II, Rabat, Maroc, Memoire de Troixieme Cycle Pour l'obtention du Diplôme d'Ingénieur d'État en Génie Rural. Available online: http://www.spate-irrigation.org/wordpress/wp-content/uploads/2011/06/MemoireNaserNajlae _GR_2006.pdf. Accessed 15 Feb 2015
PCI-Geomatics (2015) Using PCI Software, Richmond Hill, Toronto, Ontario, Canada. Available online: http://www.pcigeomatics.com/resources-support/geomatica/tutorials.pdf. Accessed 15 Feb 2015
Pike RJ, Evans IS, Hengl T (2009) Geomorphometry: A brief guide (Chapter 1). In: Tomislav H (ed) Developments in soil science: geomorphometry concepts, software, applications. and hannes, I. Reuter, vol 33. Elsevier, B.V., pp 3–30
Poudyal CP, Chang C, Oh HJ, Lee S (2010) Landslide susceptibility maps comparing frequency ratio and artificial neural networks: a case study from the Nepal Himalaya. Environ Earth Science 61:1049–1064
Quinn PF, Beven KJ, Chevallier P, Planchon O (1991) The prediction of hill slope flow paths for distributed hydrological modelling using digital terrain models. Hydrol Process 5(1):59–79
Sanyal J, Lu XX (2004) Application of remote sensing in flood management with special reference to monsoon Asia: A review. Nat Hazards 33:283–301
SIGMA (2015) Natural catastrophes and man-made disasters in 2014: convective and winter storms generate most losses. Swiss-Re-Ltd, 2, 50 (2015). Available online: http://www.biztositasiszemle.hu/files/201503/sigma2_2015_en.pdf. Accessed 10 June 2015
Toutin T (2002) Impact of terrain slope and aspect on radargrammetric DEM accuracy. ISPRS Journal of Photogrammetry and Remote Sensing 57(3):228–240
USGS (2015) Global data explorer for ASTER GDEM accessibility. Available online: http://gdex.cr.usgs.gov/gdex/. Accessed 20 Sep 2015
Vrieling A (2007) Mapping erosion from space. Ph.D. Thesis Wageningen University, Germany, ISBN 978-90-8504-587-8
Weisrock A, Wengler L, Mathieu J, Ouammou A, Fontugne M, Mercier N, Reyss JL, Valladas H, Guery P (2006) Upper Pleistocene comparative OSL, U/Th and 14C datings of sedimentary

sequences and correlative morphodynamical implications in the South-Western Anti-Atlas (Oued Noun, 29° N, Morocco). Quaternaire 17(1):45–59

Welch R, Jordan T, Lang H, Murakami H (1998) ASTER as a source for topographic data in the late 1990's. IEEE Trans Geosci Remote Sens 36(4):1282–1289

Wengler L, Weisrock A, Brochier JE, Brugal JP, Fontugne M, Magnin F, Mathieu J, Mercier N, Ouammou A, Reyss JL, Senegas F, Valladas H, Wahl L (2002) Enregistrement fluviatile et paléo-environnements au Pléistocène supérieur sur la bordure méridionale atlantique de l'Anti-Atlas (Oued Assaka, S-O marocain). Quaternaire 13(3–4):179–192

Wischmeier WH, Smith DD (1978) Predicting rainfall erosion losses—a guide to conservation planning. U.S. Department of Agriculture, Washington D.C., Agriculture Handbook, vol 537, p 58

Zhang W, Montgomery DR (1994) Digital elevation model grid size, landscape representation and hydrologic simulations. Water Resour Res 30(4):1019–1028

# Collapse Assessment of Substandard Concrete Structures for Seismic Loss Estimation of the Building Inventory in the UAE

**Aman Mwafy and Bashir Almurad**

**Abstract** Regional earthquake loss estimation systems describe the probability of losses that could happen by a seismic hazard to a certain region. In order to develop a loss estimation system for a region, the vulnerability characteristics of the exposed structures should be integrated with earthquake hazard and the inventory of the built environment. The accurate definition of structural collapse under earthquake loads is essential for deriving reliable vulnerability functions. In this study, the collapse of concrete buildings is described in terms of both global structural response and member failure, including shear failure modes. Experimentally verified shear strength models that effectively consider the reduction of shear strength with the concrete degradation under cyclic loading are implemented in a post-processor to monitor the shear supply-demand response of concrete structures under earthquake loads. A wide range of reference structures with diverse lateral force resisting systems and building heights is selected to represent substandard buildings in the UAE. Detailed fiber-based numerical models and a diverse set of earthquake records representing different seismic scenarios in the study region are employed in dynamic response simulations at various levels of ground motion intensities up to collapse. The effectiveness of the adopted shear strength models in predicting the brittle failure modes of substandard concrete buildings is demonstrated in this study. It is concluded that shear modeling is essential for the reliable earthquake loss estimation of pre-seismic code buildings. The advanced vulnerability functions confirm the need for mitigation strategies to reduce the earthquake losses of the

A. Mwafy (✉)
United Arab Emirates University, PO Box 15551, Al-Ain, UAE
e-mail: amanmwafy@uaeu.ac.ae

A. Mwafy
Faculty of Engineering, Zagazig University, Zagazig, Egypt

B. Almurad
United Arab Emirates University, PO Box 15551, Al-Ain, UAE
e-mail: 200633343@uaeu.ac.ae

 
S. Pirasteh and J. Li (eds.), *Global Changes and Natural Disaster Management: Geo-information Technologies*, DOI 10.1007/978-3-319-51844-2_14

substandard building inventory in the study area. This comprehensive study represents a step forward for the development of a reliable loss estimation system in the UAE and the surrounding region.

**Keywords** Earthquake loss estimation · Vulnerability assessment · Substandard buildings · Shear failure · Inelastic dynamic simulations · UAE

# 1 Introduction

The pre-seismic code reinforced concrete (RC) structures represented in the building inventory of the UAE need a focused attention. Although the UAE is considered a low-to-medium seismicity region as per recent seismic hazardstudies, old buildings designed to resist gravity, and wind loads only may be vulnerable to earthquake losses due to the lack of adequate seismic provisions at the construction time (Mwafy et al. 2006). This is particularly true in the light of the continuous updates in seismic hazard maps for the region. A systematic seismic vulnerabilityassessment is highly needed to predict the performance of such structures under the effect of the earthquake scenarios anticipated in the studied area (i.e., UAE). The realistic definitions of performance criteria and brittle failure modes are essential for assessing the vulnerability of substandard buildings.

Shear failureof RC structures implies rapid strength degradation and loss of energy dissipation capacity (Mwafy and Elnashai 2008). Therefore, monitoringthe shear response of RC structures, particularly substandard buildings, could be inevitable for the accurate assessment of their seismic losses. This reflects the pressing need for efficient approaches that enable tracing brittle shear failure modes in RC structures using experimentally verified shear strength models. This study thus focuses on the vulnerability assessmentof sample buildings that represent substandard RC structures in the UAE, taking into consideration the reliable assessment of shear response under the effect of different earthquake scenarios anticipated in the study region. The shear prediction models that have proven experimentally to account for the impacts of shear-axial interaction and ductility on shear strength are considered with an emphasis on substandard RC buildings.

# 2 Reference Structures and Ground Motions

## *2.1 Selection, Design, and Modeling of Pre-seismic Code Structures*

This study focuses on the vulnerability assessment of the pre-seismic code building inventory in a highly populated and active seismic area in the UAE (i.e., Dubai, Sharjah, and Ajman). The reference buildings of the present study are selected based on a building database collected for the study area (Mwafy 2012; Mwafy et al. 2015). On-ground surveys and site visits were conducted to collect the building database with the help of high-resolution satellite images. The collected structures were categorized according to four criteria, namely the building height, function, construction date, and population intensity. The buildings constructed before 1991 are considered as pre-code structures since a large percentage of these structures were designed to resist gravity and wind loads only. According to the inventory mentioned above, five RC buildings of 2, 8, 18, 26, and 40 stories were selected and designed (Issa and Mwafy 2014). Table 1 summarizes the general details of the selected reference buildings while Fig. 1 depicts the buildings classification in the study area according to their construction date. The selected 2- and 8-story buildings are frame buildings, while the 18-, 26- and 40-story buildings are shear wall structures. The analysis and design of the reference buildings were carried out using the British Standards and the ETABS structural analysis software (BS8110 1986; CSI 2011). The material properties were selected to represent those expected at the time of construction (Issa and Mwafy 2014). The required amount of longitudinal reinforcement was calculated for the columns, walls and core walls of the reference buildings. The design shear strength is estimated and compared with the shear demand to calculate the transverse reinforcement.

Detailed modeling of the reference structures was conducted for inelastic analysis (Issa and Mwafy 2014). The numerical models were developed using the inelastic analysis platform ZEUS-NL (Elnashai et al. 2012). The materials were selected to idealize effectively the reinforcing steel bars, confined, and unconfined concrete during the inelastic simulations. For the 40-, 26- and 18-story structures, the reference buildings were idealized as two-dimensional (2D) systems. The frame structures (i.e., 2- and 8-story buildings) were idealized as three-dimensional (3D) models. The finite element models (design) and fiber-based (assessment) models of all reference buildings are depicted in Fig. 2.

**Table 1** Details of the reference buildings (Issa and Mwafy 2014)

| Number | Building reference | No. of stories | Story height (m) | | | Total height (m) |
|---|---|---|---|---|---|---|
| | | | B | GF | TF | |
| 1 | BO-02 | 2 | – | 5.0 | 3.5 | 8.5 |
| 2 | BO-08 | 8 | – | 5.0 | 3.5 | 28.5 |
| 3 | BO-18 | 18 | 3.2 | 4.5 | 3.2 | 58.9 |
| 4 | BO-26 | 26 | 3.2 | 4.5 | 3.2 | 84.5 |
| 5 | BO-40 | 40 | 3.2 | 4.5 | 3.2 | 129.3 |

*B* basement; *GF* ground floor; *TF* typical floor

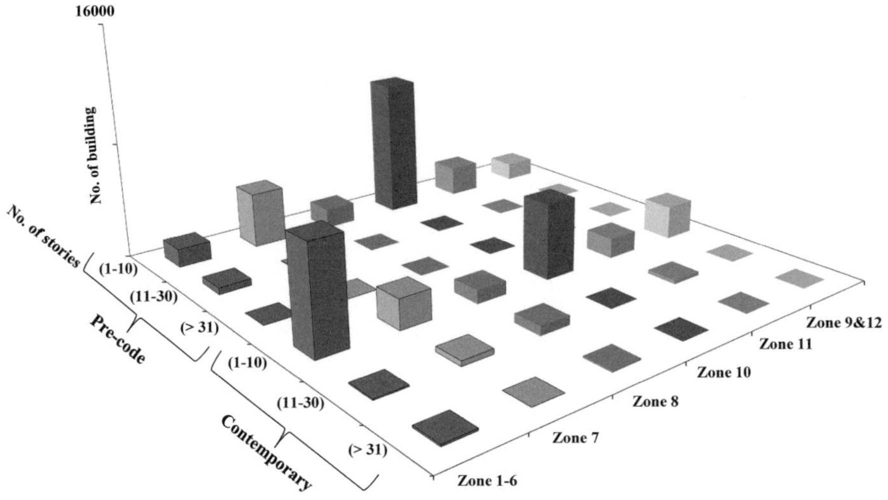


**Fig. 1** Buildings classification according to their construction date (Issa and Mwafy 2014)

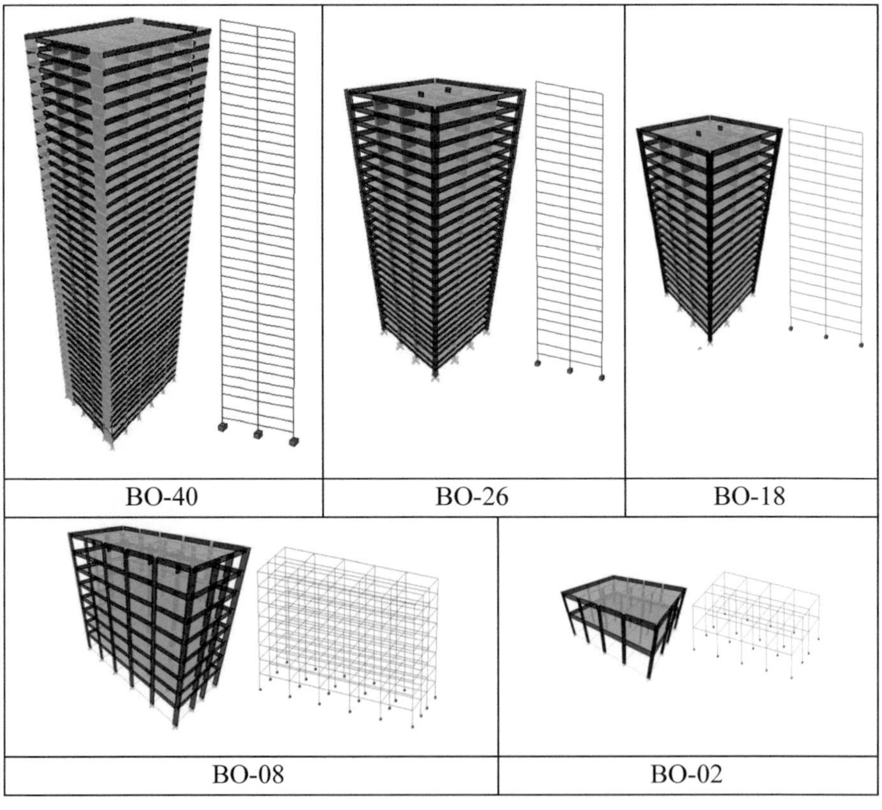


**Fig. 2** Finite element and fiber-based models of reference structures

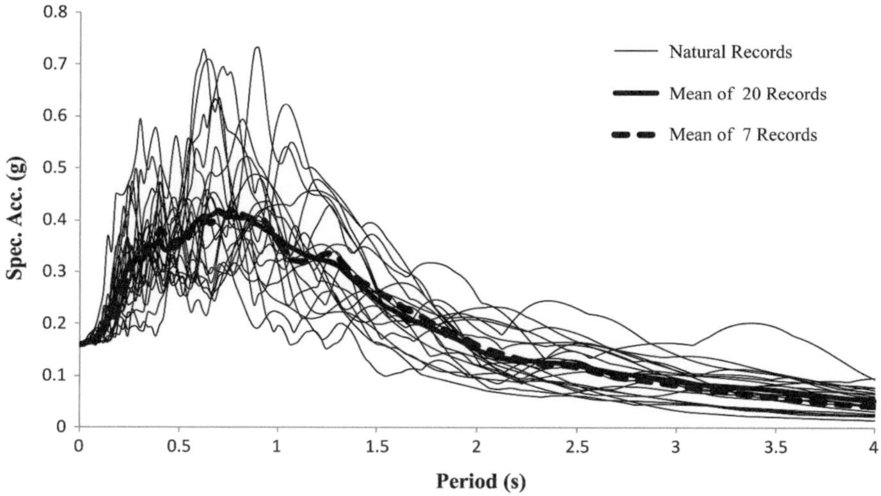


**Fig. 3** Comparison of mean response spectra of two groups of 20 records and seven events representing far-field earthquake scenarios

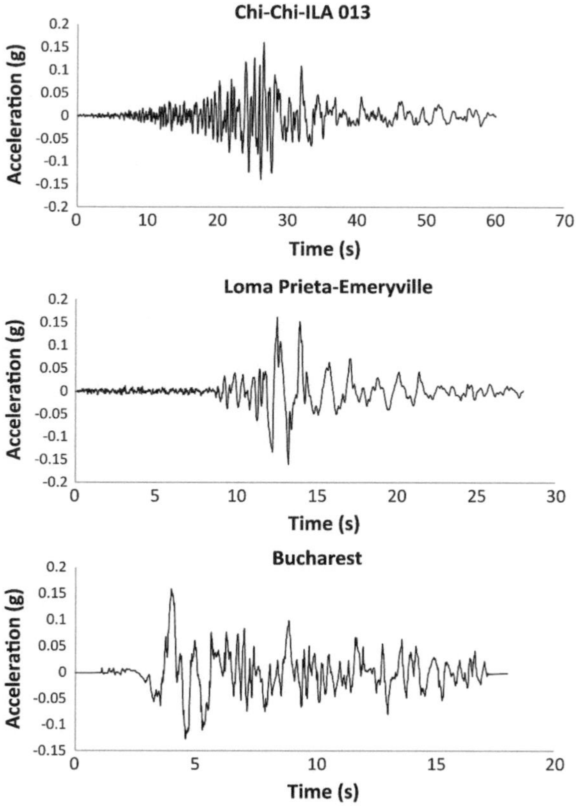


**Fig. 4** Sample of the selected far-field earthquake records

### 2.2 Selection of Scenario-Based Earthquake Records

Two earthquake scenarios representing far-field and near-source earthquake records are adopted in this study based on the recommendations of previous seismic hazard studies for the UAE (Mwafy et al. 2006). Fourteen far-field and near-source records are selected to represent the two seismic scenarios mentioned above. The mean of the selected records matches the mean of 40 earthquake records used in previous vulnerability assessment studies for the UAE (Mwafy 2012; Mwafy et al. 2015). The response spectra of 20 far-field earthquake records along with their mean are compared with the mean of selected seven records in Fig. 3. Samples of the selected far-field earthquake records are also shown in Fig. 4.

## 3 Performance Assessment of Existing Structures

### 3.1 Shear Demand-Supply Response at the Member Level

A literature review is carried out to select experimentally verified shear strength models in addition to the code design approach for the assessment of the shear response of RC columns and shear walls. Five shear strength models for RC columns are selected, namely those proposed by Kowalsky and Priestley 2000; Priestley et al. 1994; Sezen and Moehle 2004; Kim et al. 2012; Howser et al. 2010; Bentz et al. 2006. The latter shear strength approach is based on the simplified modified compression field theory (SMCFT). Three shear strength models for walls are selected, namely those recommended by Kowalsky and Priestley 2000; Priestley et al. 1994; Wallace 2010; Krolicki et al. 2011. The design provisions currently implemented in the reference study area is also used in the shear performance assessment (ACI-318 2011). The performance assessment of the reference structures is conducted using both the inelastic pushover analysis (IPA) and time history analysis (THA). The response of each of the investigated buildings is presented at three critical story levels, namely at the basement level, building mid-height and the roof of each structure. The shear response is traced, and any indications of shear failurein structural members are reported when the shear demand exceeds the capacity.

The structural performance is verified using IPA by applying lateral load along the building height (Mwafy and Elnashai 2001). The shear supply is compared with the shear demand of structural member at different inter-story drift ratios (IDRs). Sample results for the shear response of the 18-story building are presented in Fig. 5. Shear failureis observed in the core walls and external shear walls at the basement level as per the Wallace (2010) shear strength model. The axial force increases with increasing lateral loads, and the shear demand exceeds the supply

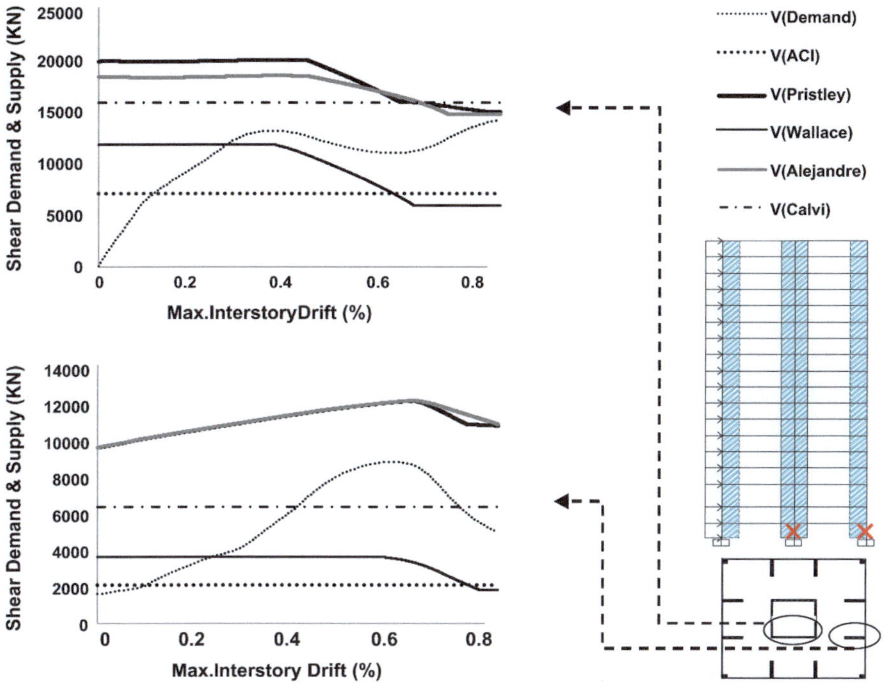


**Fig. 5** Shear demand-supply response of core walls and external shear walls at the basement level using IPA (18-story building)

causing shear failure. The negative impact of ductility demands on the shear supply under high levels of lateral loads and IDRs is shown from the shear response predicted using the Priestley (Kowalsky and Priestley 2000), Calvi (Krolicki et al. 2011) and Wallace (2010) shear strength models.

Furthermore, the shear response of the reference structures is monitored under the effect of two earthquake scenarios at different intensity levels. Sample results are presented in Fig. 6 for the 40-story building under the near-source earthquake record Lazio Abr. It is shown that the core walls at the basement level fail in shear at high PGA levels. All of the selected shear strength models detect shear failure. Under near-source earthquake records, the shear response is significantly influenced by the variation of axial loads, without observable effects from ductility.

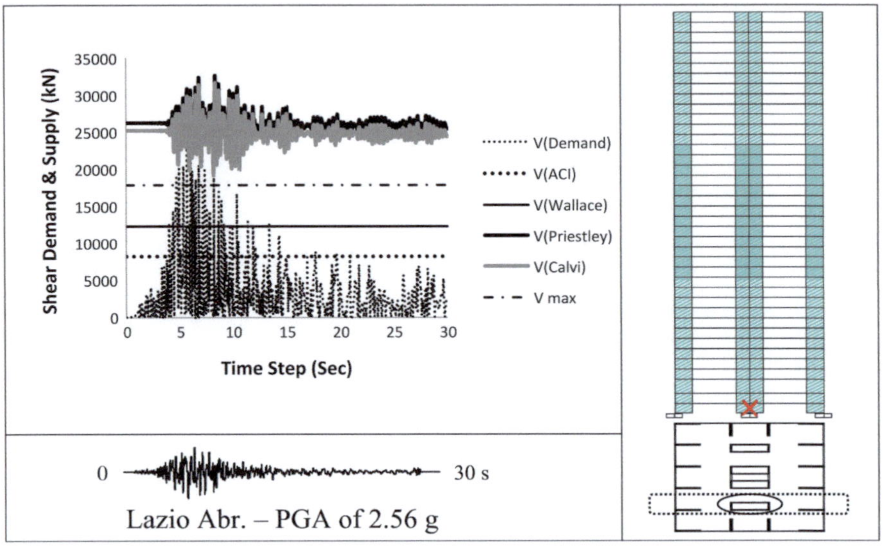


**Fig. 6** Shear demand-supply response of the core walls of the 40-story building at the basement level under a near-field earthquake record

## 4 Selection of Performance Indicators

Three performance limit state criteria are adopted for fragility analysis, namely immediate occupancy (IO), life safety (LS), and collapse prevention (CP). The IO limit state defines a minor damage state in which the building remains safe to occupy after the earthquake while any needed repairs are minor. The LS limit state represents a significant damage sustained by the building while the structure still has some reserve capacity. The CP limit state defines a significant state of damage in which the building is on the verge of structuralcollapse(Jeong et al. 2012). THAs are carried out to predict the shear failure of the reference buildings, and IDRs are obtained once the shear failureis detected from different seismic scenarios. The average IDRs when shear failure is detected in the five reference buildings are presented in Fig. 7 for the far-field earthquake records. The comprehensive results obtained from a large number of far-field and near-source earthquake records indicate that the average IDR values at shear failureobtained from the near-source records are less than those from far-field events for the shear wall structures. On the other hand, the average IDRs of the frame structures at shear failure increase under the effect of near-source records when compared with the far-field counterparts. The results suggest adopting seismic scenario-based limit state criteria to quantify the level of damage to different structural systems.

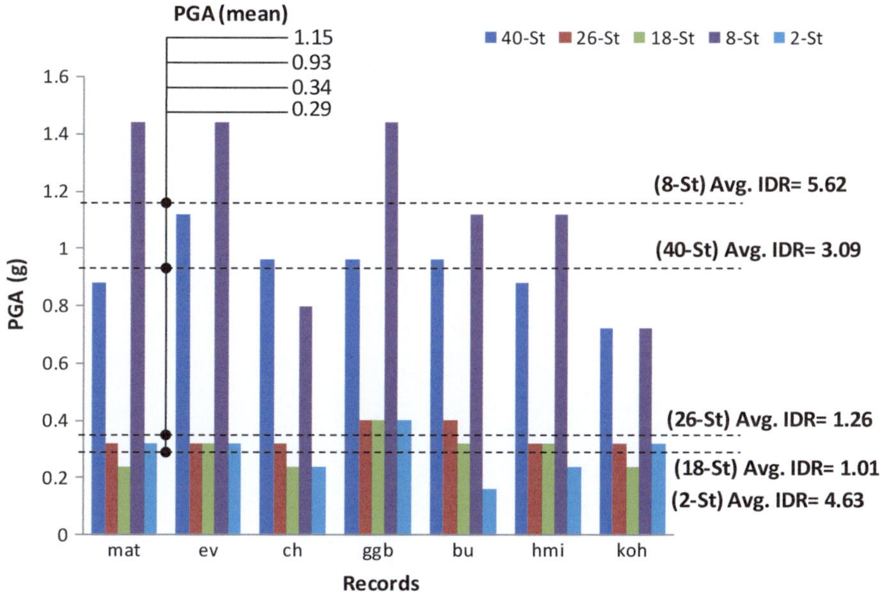


**Fig. 7** Average IDRs of five reference buildings under the effect of seven far-field earthquake records

The performance criteria of the reference structures were selected in a previous vulnerability assessment study without estimating the impact of shear response (Issa and Mwafy 2014; Mwafy et al. 2015). In the present study, limit state criteria are selected according to the THA results considering the effect of shear response from different earthquake scenarios on limit states. For the 40-story building, shear failureis detected earlier under the near-source earthquake records compared with the far-field counterparts, as shown in Table 2. The impact of shear assessment on the results under the effect of far-field records is insignificant. For the 26- and 18-story buildings, the effects of shear response on the results obtained from both near-source and far-field earthquake records are observable, as shown in Table 2. Revised CP limit states are therefore selected for both the far-field and near-source earthquake events. On the other hand, the effect of shear assessment on the results of frame structures under both the far-field and near-source records is insignificant, and hence no changes in limit states are adopted for this class of structures. The adopted limit states in the current study are summarized in Table 2. It is noteworthy that the LS limit state is considered as 50% of the CP performance criterion (ASCE/SEI-41 2007).

**Table 2** Revised limit states of five reference buildings using different earthquake scenarios considering shear assessment

| Earthquake scenario | Reference structure | | | | | | | | | | | | | | |
|---|---|---|---|---|---|---|---|---|---|---|---|---|---|---|---|
| | 40-st | | | 26-st | | | 18-st | | | 8-st | | | 2-st | | |
| | Limit state—inter-story drift (%) | | | | | | | | | | | | | | |
| | IO | LS | CP[a] | IO | LS | CP[a] | IO | LS | CP[a] | IO | LS | CP[a] | IO | LS | CP[a] |
| Far-field records | 0.34 | 0.89 | 1.78 | 0.34 | 0.63 | 1.26 | 0.34 | 0.505 | 1.01 | 0.39 | 1.48 | 2.96 | 0.39 | 1.48 | 2.96 |
| Near-source records | 0.34 | 0.65 | 1.3 | 0.34 | | 0.6 | 0.34 | | 0.47 | | | | | | |

[a]Excluding shear assessment, the IDR corresponding to the CP limit state was 1.78% for the 18–40 story buildings, while it was 2.96% for the 2- and 8-story structures

## 5 Fragility Assessment

Fragility curves represent the probability of exceeding limit states at different ground motion intensity levels (Jeong et al. 2012). The expected physical damage of a structure can be determined from fragility curves. The vulnerability functions can be therefore integrated with loss estimation systems to provide predictions of earthquake losses. Taking into consideration the revised limit states of the present study, as presented in Table 2, improved fragility curves are developed and compared with those derived in previous studies (Issa and Mwafy 2014). Since shear assessment has observable impacts on the performance of shear wall structures, revised fragility curves are only derived for this group of structures. The fragility curves of the frame structures (i.e., 2- and 8-story buildings), which were previously developed (Issa and Mwafy 2014), are unaffected by the shear assessment conducted in the present study, as shown in Fig. 8a for the 8-story building. For the 18-story building, the limit state exceedance probabilities significantly increase under the far-field earthquake records using the revised performance limit states, as presented in Fig. 8b. For the 26-story building, the limit state exceedance probabilities also increase, but with less extent, under far-field earthquake records using the revised performance limit states, as shown in Fig. 8c. The steepness of the CP and LS fragilities increases for both the 26- and 18-story buildings, while the

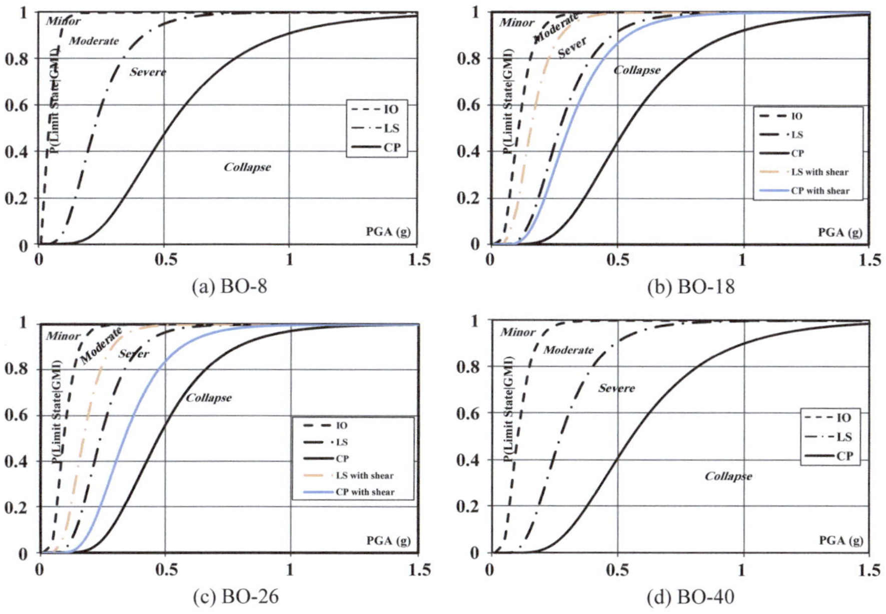


**Fig. 8** Fragility curves of four reference structures using a far-field earthquake scenario

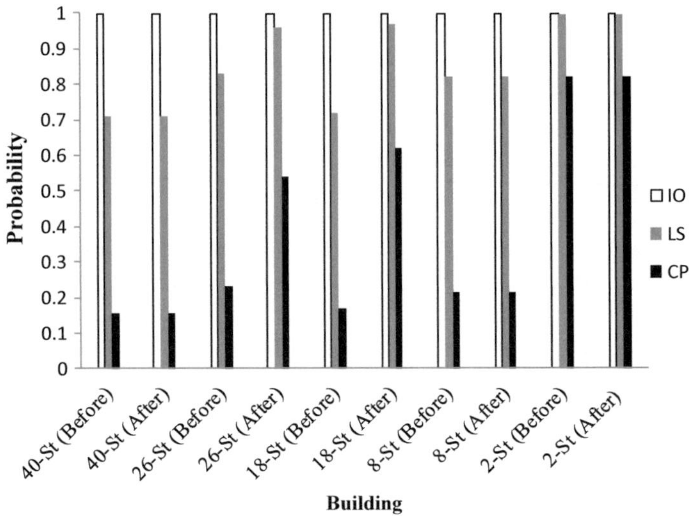


**Fig. 9** Limit state exceedance probabilities for far-field earthquake records at twice the design PGA

fragility curves of the 40-story building under the far-field earthquake records remain as recommended in previous studies without any changes (Issa and Mwafy 2014; Mwafy et al. 2015), as shown in Fig. 8d.

The probability of exceeding the revised performance limit states clearly describe the impact of shear response on the probabilistic assessment study carried out for the five reference structures. Figure 9 depicts a comparison between the limit state exceedance probabilities obtained from the far-field earthquake records at twice the design PGA with and without the shear assessment of the reference buildings. The probability of damage significantly increases for the CP and LS limit states of the 26- and 18-story buildings. The results clearly suggest that the CP and LS damage probabilities increase as the height of structures decreases.

## 6 Conclusions

This study focused on the earthquake vulnerability assessment of frame and shear wall buildings that represent substandard RC structures in the UAE, taking into consideration the reliable assessment of failure modes. Five reference RC buildings of 2, 8, 18, 26, and 40 stories were selected, fully designed and idealized for dynamic response simulations. Two seismic scenarios representing far-field and near-source earthquake records were adopted to represent the expected earthquake scenarios in the study region. Several experimentally verified shear strength models for RC columns and shear walls were selected and implemented with other structural performance indicators in a post-processor to enable the reliable assessment of the reference buildings under earthquake loading.

The presented sample results from inelastic pushover analyses and dynamic response simulations indicated that shear failure was detected in the pre-code shear wall structures unlike the case of frame buildings. For the 40-story shear wall building, the impact of shear assessment on the performance limit states was insignificant under the effect of far-field records, while the brittle shear failure was detected much earlier under the near-source earthquake records. For the 26- and 18-story buildings, the effects of shear response on the results obtained from both near-source and far-field earthquake records were observable, and hence revised limit states were selected for both earthquake scenarios. Seismic scenario-based performance limit states were therefore adopted based on the comprehensive results of this study to quantify the level of damage to different structural systems. Considering the proposed performance limit states, improved fragility relationships were developed for pre-code buildings. The vulnerability relationships indicated that the limit state exceedance probabilities significantly increase for the CP and LS performance criteria, particularly for medium-rise shear wall structures (i.e., 18- and 26-story buildings). The results clearly confirmed that the damage probabilities increase with decreasing the building height. This earthquake vulnerability assessment study for the pre-code buildings in the UAE contributes in developing a holistic earthquake risk management system for the building and infrastructure in the UAE and the surrounding region.

**Acknowledgments** This work was supported by the United Arab Emirates University under research grants No. 31N227 and 31N132.

## References

ACI-318 (2011) Building code requirements for structural concrete and commentary. American Concrete Institute, Detroit

ASCE/SEI-41 (2007) Seismic rehabilitation of existing buildings, standard ASCE/SEI 41-06. American Society of Civil Engineers, Reston

Bentz EC, Vecchio FJ, Collins MP (2006) Simplified modified compression field theory for calculating shear strength of reinforced concrete elements. ACI Struct J 103:614

BS8110 (1986) Structural use of concrete. British Standard Institution, London

CSI (2011) ETABS—integrated building design software. Computers and Structures Inc., Berkeley

Elnashai AS, Papanikolaou V, Lee D (2012) Zeus-NL—a system for inelastic analysis of structures—user manual. University of Illinois at Urbana-Champaign, Urbana

Howser R, Laskar A, Mo Y (2010) Seismic interaction of flexural ductility and shear capacity in reinforced concrete columns. Struct Eng Mech 35:593–616

Issa A, Mwafy AM (2014) Fragility assessment of pre-seismic code buildings and emergency facilities in the UAE. In: Second European conference on earthquake engineering and seismology (2ECEES), Istanbul, Turkey, 24–29 Aug 2014

Jeong SH, Mwafy AM, Elnashai A (2012) Probabilistic seismic performance assessment of code-compliant multi-story RC buildings. Eng Struct 34:527–537

Kim I-H, Sun C-H, Shin M (2012) Concrete contribution to initial shear strength of RC hollow bridge columns. Struct Eng Mech 41:43–65

Kowalsky MJ, Priestley MN (2000) Improved analytical model for shear strength of circular reinforced concrete columns in seismic regions. ACI Struct J 97:388–396

Krolicki J, Maffei J, Calvi G (2011) Shear strength of reinforced concrete walls subjected to cyclic loading. J Earthquake Eng 15:30–71

Mwafy AM (2012) Classification and idealization of the building stock in the UAE for earthquake loss estimation. In: 15th world conference on earthquake engineering, Lisbon, Portugal, 24–28 Sept 2012

Mwafy AM, Elnashai AS (2001) Static pushover versus dynamic collapse analysis of RC buildings. Eng Struct 23:407–424

Mwafy AM, Elnashai AM (2008) Importance of shear assessment of concrete structures detailed to different capacity design requirements. Eng Struct 30:1590–1604

Mwafy AM, Elnashai AS, Sigbjornsson R, Salama A (2006) Significance of severe distant and moderate close earthquakes on design and behavior of tall buildings. Struct Des Tall Spec Build 15:391–416

Mwafy AM, Ashri A, Issa A (2015) Probabilistic vulnerability assessment of the building inventory in an extended seismically active area in the UAE. In: 3rd international conference on engineering geophysics, Al Ain, UAE, 15–18 Nov 2015

Priestley MJN, Verma R, Xiao Y (1994) Seismic shear strength of reinforced concrete columns. J Struct Eng 120:2310–2329

Sezen H, Moehle JP (2004) Shear strength model for lightly reinforced concrete columns. J Struct Eng 130:1692–1703

Wallace JW (2010) Performance-based design of tall reinforced concrete core wall buildings. In: Earthquake engineering in Europe. Springer, London, pp 279–307

# Part VI
# Disaster Management

# Status of Spatial Analysis for Urban Emergency Management

**Rifaat Abdalla**

**Abstract** This paper provides multidisciplinary coverage for the use of geospatial information systems for urban emergency management with emphasis on selected common types of emergencies. The paper presents selected urban emergencies that are globally known and pose a high risk for considerable parts of the world. The review addresses the application issues of how spatial analysis can be utilized in addressing emergency management issues by defining its usability and limitations in dealing with the questions of vulnerability and risk assessment as well as the state of the art of approaches that can be used for risk mapping and visualization. The objective is to provide conceptual coverage to timely solutions for emergency preparedness and response. The paper emphasized that among issues that may face accurate utilization of spatial analysis, is the accuracy of data and time of processing, as well as shared coordination among stakeholders. This research concludes that a challenge to effective risk reduction is providing disaster managers with access to data and approaches that may help them in analyzing, assessing, and mapping risk models.

**Keywords** GIS · Emergency management · Urban centers · Spatial analysis · Environmental modeling

## 1 Introduction

The applications of GIS in Disaster Management are increasingly becoming an integral element of disaster and emergency management activities globally. The time factor is very critical in emergency management operations, and it demands decision-makers to make swift decisions under time pressure. The spatial nature of critical information makes spatial data to be very crucial in decision-making

R. Abdalla (✉)
Department of Hydrographic Surveying, Faculty of Maritime Studies,
King Abdulaziz University, P. O. Box 80401,
Jeddah 21589, Kingdom of Saudi Arabia
e-mail: rmabdalla@kau.edu.sa

 
S. Pirasteh and J. Li (eds.), *Global Changes and Natural Disaster Management: Geo-information Technologies*, DOI 10.1007/978-3-319-51844-2_15

operations, for disaster management. It is imperative for decision-makers to access the right data at the right time presented inaccurate models to allow them to respond, plan, or mitigate disasters. The time versatile of emergency do not permit decision-makers to gather the necessary information. Thus, most of the time, pre-planned emergency management scenarios are used (Becerra-Fernandez et al. 2008). GIS Technology can fill the gap of visualization and analysis of what-if scenarios. This allows emergency managers to have access to adequate information stored in spatial databases and presented in a computer-generated maps or interactive models (Miura et al. 2007). GIS can be very helpful to make a well-thought plan for emergency response operations, and it can address the general public. It is a training, very useful tool in emergency management table top exercises, and an integral part of Emergency Operations Centers (EOC) (ESRI 1999). GIS provides a mechanism for visualization and modeling of critical information at a single or multiple locations during an emergency (ESRI 1999). This provides a user-driven framework, which encompasses the phases of disaster management, to support the process of enhanced decision-making and increases the degree of involvement of each team of personnel associated activities and analytical procedures (Smirnov et al. 2006).

When discussing emergency management, mostly we are addressing the issues of preparedness. This is a fundamental component of disaster and emergency management and can play a vital role if emergency response actions become essential. The usefulness of GIS as a decision-making tool in helping emergency management decision-makers falls in the following:

1. Assessing risks
2. Evaluating threats
3. Tracking what-if scenarios
4. Maintaining situational awareness
5. Documenting disparity
6. Ensuring the focused allocation of resources
7. Alerting and notifying communities
8. Minimizing the disruption caused by necessary interventions during the response phase. Figure 1 shows the elements of GIS in Urban Emergency Management.

The factors above are particularly of importance from spatial analysis perspective (Saadatseresht et al. 2009) In an emergency situation, spatial analysis can be run for evacuation, to relocate public from a hazardous region to a safer region, and it usually times critical activity (Anjum et al. 2011). It is necessary to utilize the functionality of spatial analysis tools to formulate evacuation strategies to have a good response in an emergency state. A central challenge for using spatial analysis in developing an evacuation plan is in determining the disruption of services and in the distribution of evacuees into the safe areas. Precisely, supporting the decision of where and from which road each evacuee should go is a key functionality of spatial analysis (Cova and Church 1997). To effectively achieve the aim of spatial analysis

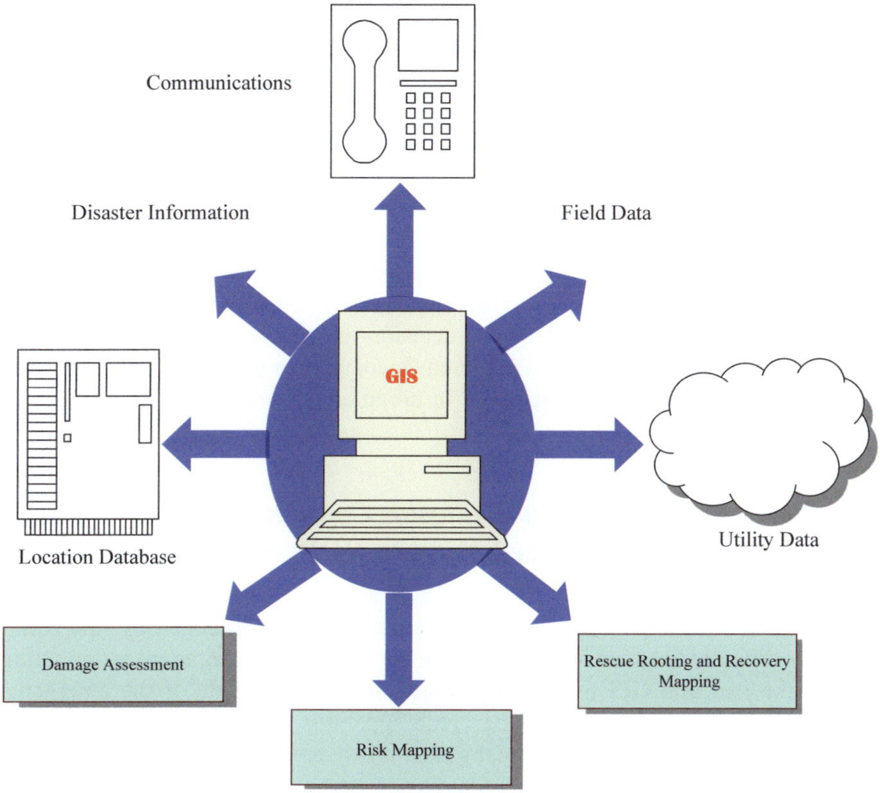


**Fig. 1** Various GIS emergency management elements

in emergency management operations, several objectives are brought into consideration and satisfied simultaneously through this paper, nevertheless, these impartial roles may repeatedly encounter with each other. These goals are: (A) How a decision-maker can utilize the functionality of spatial analysis to make critical decisions during emergencies? (B) What are essential functions that spatial analysis can help with during emergency management cycle in the ten selected emergency types? (C) How emergency managers could better implement spatial analysis operations as part of their day-to-day operations?

Effective disaster management calls for involving multimodal decision support capabilities, that involves groups at all levels of forecasting and response, in addition to collective mitigation efforts that address the origin of vulnerability (Morrow 1999). The vulnerability of community is directly attributed to the socioeconomic factors that affect the community, i.e., it directly impacted, whether increased or decreased as a result of the socioeconomic well-being of a community, as it relates to their day-to-day activities (ESRI 1999). Emergency planners, policy-makers, risk analysts, and first responders usually try to classify and locate high-risk sectors using the tool known as Community Vulnerability Maps, incorporating this data into GIS systems where practicable, and for this, spatial analysis is crucial (Kumar 2013).

## 2 Overview of Urban Emergency Situations

This section provides a critical review on the usability of spatial analysis in selected urban emergency situations. It provides an in-depth coverage of the work cited in this regard with emphasis on enhanced decision-making process. The scope of the coverage will focus on the most notable advancements in the use of spatial analysis techniques for emergency management in urban environments.

### *2.1 Earthquakes and Humanitarian Coordination for Internally Displaced Persons (IDP)*

The literature review on GIS and humanitarian coordination began by first examining the different ways in which GIS can be applied humanitarian coordination. Despite the fact that GIS has been predominately seen, inside of the emergency management community, as a cartographic apparatus, a way to deal with critical thinking, or an electronic navigational guide, this does not satisfactorily portray the state-of-the-art use of GIS in humanitarian assistance (Currion 2006). There are many potential applications of GIS for humanitarian aid. For example, the application of optimization, which is the use of advanced GIS algorithms to solve a design problem, can be used to find suitable locations for evacuation. For instance, a buffer analysis, which is a GIS transformation tool, can be used to estimate vulnerability based on proximity to different hazards.

In 2005, a Complex Humanitarian Emergencies Study (Verjee 2005) drew from case studies and patterns in technological advancement to layout the potential GIS applications for humanitarian emergencies, which were:

1. Mapping and Cartography (Land use Mapping, Infrastructure Mapping, demographic Mapping, Logistics, and Sustainability).
2. Outreach, Media, and Communications (Public information, Reporting, Program Assessment, News coverage).
3. Modeling and Simulation for Disaster Scenarios (rehearsal, Data and information flow patterns, planning for contingencies).
4. Environmental Management and Planning (Yield cultivation, resources assessment, and planning).
5. Risk and Hazard Management (seismic analysis, site location planning, and water-level measurement and mitigation).

6. Vulnerability Analysis and Assessment (Early warning systems for drought, desertification and famine, Epidemics modeling, and Tsunami Planning).
7. Risk Reduction ("hot spots" identification and mitigation programming).
8. Response Policies and Organizational Management (management, planning, and training) (Verjee 2005).

Table 1 is showing the application of GIS in this situation. Although there are numerous applications of spatial analysis as a GIS technique, they all share the ultimate objective, which is to make the most of the situational awareness of all specialists participating so that main concerns can be recognized and then jointly attained.

**Table 1** Applications of spatial analysis in humanitarian coordination modified Abdalla (2016)

| GIS application | Description of application | Tools | Application examples |
|---|---|---|---|
| *Queries and measurements* | | | |
| Queries | Does not change the database or produce new data but reveals information | Scatter plots, residuals, and structured/standard query language | To determine water sources near potential IDP settlements |
| Measurements | Numerical interrogations of GIS, which make an analysis of spatial data | Makeup distance, area, length, shape, aspect of spatial data | To estimate the total area needed to set up IDP settlements<br>To determine the length between food distribution points |
| *Transformations* | | | |
| Transformations | Creates new data from existing data | Buffer analysis | Used to comply with Sphere humanitarian standards. For example, can be used to ensure that toilets are within 50 m of shelters |
| Spatial interpolation | A transformation analysis method used for intelligently guessing the value of the discreet object | Thiessen polygons, inverse-distance weighting, and kriging | To predict infection rates in IDP settlements<br>To estimate flow rates of well to large aquifer |
| *Optimization* | | | |
| Point optimization | Determines optimal locations amongst nodes of a network | | To select location for critical infrastructure such as food storage areas |
| Route optimization | Determines optimal routes amongst nodes of a network | | To access accessibility of road networks to potential IDP settlements |

(continued)

**Table 1** (continued)

| GIS application | Description of application | Tools | Application examples |
|---|---|---|---|
| Path optimization | Solves are routing problems by minimizing friction value and does not require a network | | To determine paths for utility infrastructure for IDP settlements |
| *Analysis* | | | |
| Geostatistical analysis | Used to measure geographic distributions, identify pattern and clusters, and analyze geographic relationships | | To determine historical rainfall averages over a particular area, which could be applied to potential IDP locations |
| Centroid analysis | Used to identify trends over various phenomena in a set period | | To determine trends in infectious diseases and household income levels within IDP settlements |
| Pattern analysis | Used to identify distribution points. Can determine whether the point is random, clustered, or dispersed | | To identify links between water wells and infectious diseases |
| Relationship analysis | Used to determine the relationship between various geographic phenomena | | To assist in increasing the efficiency of daily operation of IDP settlements |
| Geovisualization | Simple three dimensions (3D) and high-resolution mapping are used globally to access a variety of information | Google Earth, ESRI's ArcExplorer, and Microsoft Virtual Earth | Allows a layperson to understand the unfolding of humanitarian emergency |

A recent analysis by Eveleigh et al. (2007) acknowledges, in the context humanitarian emergencies "GIS technology is struggling with how to address complex problems that require the modeling of dynamic phenomena, feature behavior, and process information". Eveleigh et al. (2007) conclude that there is a strong potential for GIS-based analysis models to provide the breakthrough needed to address the complex nature of humanitarian emergencies. Figure 2 is showing some of the impact of earthquake in populated areas.

The (Bally et al. 2005) 'Indicated that Remote Sensing and Humanitarian Aid provide a life-saving combination'. The use of remote sensing and GIS allowed 200,000 IDPS to be relocated to longer term settlements that had renewable water reserves as well as development potential regarding sanitation, agriculture, and even hydroelectricity (Bally et al. 2005). Another influential GIS application used to support humanitarian emergencies was The Global Connection Project, which involved Carnegie Mellon University, NASA, Google, and National Geographic, contributing to the relief planning for the October 8, 2005, South Asian earthquake.

**Fig. 2** IDP settlement during Nepal earthquake of 2015, photo credit OCHA

In this project, GIS was used to acquire and deliver high-resolution imagery from Digital Globe's Quickbird.

## 2.2 *Urban Fire*

To accurately portray a wildfire situation, spatial analysis can be used to identify high-risk fire zones and establish buffer areas (ESRI 1999). In addition to the identification of high-risk areas, spatial analysis can be combined with statistical analysis as a verification method for estimation of the concentration of population in different zones as well as for determining damage assessment models for the region (Goodchild 2006). Pew and Larsen (2001) provided guidance by identifying potential layers that can be used for urban fire detection. Our first step is to use latitude and longitude coordinates to plot the various fires (based upon a choice of lightning or human-ignited fires) within a given period. Fire data may appear to be located within bodies of water, but this is simply a cause of rounding error and buffer zones provide some leeway for such error. The process of linking attributes data and present four analytical methods for wildfire. Although they focus specifically on human-caused fire disasters, in suggesting the four recommended options for complete urban fire analysis:

1. the area affected
2. temporal variation
3. spatial variation
4. probability.

The urban fire hazard is difficult to prevent. However, through the identification of the high-risk zones, the frequency of fire can be minimized. This research indicates that GIS when combined with satellite imagery can identify high-risk areas within a given proximity and grade the level of sensitivity to fire about surrounding areas. Jaiswal et al. (2002) discuss the use of ArcGIS for this concept, asserting that the combination of topographic infrastructure information and remote sensing for vegetation mapping can create an accurate estimation of high-risk fire areas used for mitigation and response purposes. In Jaiswal et al. (2002) layers of vegetation types, slope, proximity to settlements, and distance from roads were created about high-risk fire areas (Jaiswal et al. 2002). After plotting this data, buffer zones of 1000, 2000, 3000, and 4000 m surrounding the high-risk areas were plotted to project the different levels of danger (Jaiswal et al. 2002).

Although Jaiswal et al. (2002) looked specifically at the case of India, this concept of GIS combined with satellite imagery for identification of high-risk areas proved the effectiveness of GIS as a tool for disaster management. If GIS can be used to map high-risk zones with a buffer area, which provides baseline understanding that GIS could also be used to demonstrate damage assessment models using ArcGIS.

Pradhan et al. (2007) used GIS analysis to determine fire susceptibility, using a "vector-type spatial database" with GIS and combined with topographic data, fuel data, base survey points, and maps. This allowed for calculating factors, which were then converted to a raster grid, identifying 112 cells within the fire occurrences (Pradhan et al. 2007). A frequency-based ratio approach was adopted to define the "relationships between hotspot locations and the factors in the study area" (Pradhan et al. 2007). The difficulties, however, were in processing "a significant amount of data" (Pradhan et al. 2007). The conclusion drawn from Pradhan et al. (2007) regarding the use of such projections for mitigation purposes was of particular interest. In predicting fire susceptibility when using frequency analysis, the authors recommend using the results with caution. The authors recommend that their analysis be used primarily during fire situations, which suggests mapping fire-affected areas rather than leading toward the mitigation portion of fire disaster management.

## 2.3 Terrorists Attacks

Kwan and Lee (2005) displayed that terrorist attacks at the World Trade Center (WTC) in New York City and the Pentagon on September 11, 2001, have not just influenced multilevel structures in urban ranges additionally affected by their quick surroundings at the road level in ways that impressively decreased the velocity of crisis reaction. The capability of utilizing ongoing 3D GIS for the advancement and execution of GIS-based clever crisis reaction frameworks (GIERS) was analyzed. The point was at encouraging snappy crisis reaction to terrorist assaults on multilevel structures (e.g., multistorey office structures). A framework engineering and a system information display that coordinates the ground transportation framework with the interior courses inside multilevel structures into a safe 3D GIS was inspected. Issues of utilizing versatile representation stages were additionally talked about particularly the requirement for the remote and portable arrangement. Key choice bolster functionalities of GIERS were additionally investigated with specific reference to the use of system-based most limited way calculations. A test usage of a coordinated 3D system information model utilizing a GIS database for a close study range was displayed by Kwan and Lee (2005). The study shows that response delay within multilevel structures can be much longer than delays incurred on the ground transportation system, have the potential for considerably reducing these delays.

Johnson (2003) indicated that in times of emergency, the emergency management decision-makers have the essential obligation regarding rapidly and effectively dealing with any situation that may happen. An altered GIS application was produced empowering a fleeting-based investigation of a disaster event coordinated with the centralization of populaces identified precisely to the room level. The GIS Emergency Management System (GEMS) application is an intelligent framework to be used in the Emergency Operation Center to boost the bearing of the reaction. If a disaster has to happen, the mediation and recuperation endeavors could be at first centered on the most basic regions with the biggest concentration of individuals.

## 2.4 Infrastructure Failure

Cova and Church (1997) presented a technique for deliberately distinguishing neighborhoods that may confront transportation challenges during the evacuation. A characterization of this nature offers an interesting way to deal with evaluating group helplessness in locales subject to quick moving risks of indeterminate spatial effect (e.g., dangerous spills on roadways). A heuristic calculation is depicted which is fit for delivering productive, high-quality solution for this model in a GIS setting, as it was connected to a study area.

A new computerized risk management system for use by personnel who are not risk experts may reduce the probability and seriousness of accidents (Camps 1993).

The system, which was created is appropriate for use in oil, gas, or chemical processing sites. It incorporates mathematical models, calculation tools for accident simulation, and a database that includes accident scenarios. It can also be used in an emergency situation to determine preferred ways to find external assistance.

## 2.5 Flood Scenarios

Correia et al. (1998) showed that GIS has been perceived as an effective intends to coordinate and examine information from different sources with regard to far-reaching floodplain administration. As a component of this worldwide way to deal with floodplain administration, it is vital to have the capacity to foresee the results of various situations in regard to overflowed territories and related danger. Hydrologic and water-powered displaying assume a fundamental part, and there is much to pick up in consolidating these demonstrating abilities in GIS. Johnson (2003) presents results in view of the utilization of Intergraph GIS combined with IDRISI GIS. Utilizing these two frameworks generously expanded the adaptability of utilizing GIS as an instrument for surge concentrates on. As the hydrodynamic reproduction is straightforwardly consolidated with a GIS, the outcomes can be effortlessly handled as neighborhood immersion profundities for spatial danger dissects.

The role of GIS in flood emergency management was examined by Cova (1999), through the viewpoint of complete crisis administration (CEM) and its four stages: moderation, readiness, reaction, and recuperation. In the wake of a calamity, GIS is getting to be necessary for supporting harm appraisal, modifying, and state-funded training. The part of GIS in surge crisis administration was inspected by Cova (1999), according to the analysis of far-reaching crisis administration (CEM) and its four stages: alleviation, readiness, reaction, and recuperation. In the wake of a catastrophe, GIS is getting to be necessary for supporting harm appraisal, reconstructing, and government-funded instruction. Abbas et al. (2009) proposed a GIS-construct study on the improvement of surge displaying and representation for Allahabad Sadar Sub-District (India) be exhibited. This incorporates the outline, the procedure/approach that intended to investigate the degree for spatial examination application for catastrophe administration. The surge-inclined zones have been distinguished, and their positions are checked, where the GIS usefulness has been abused to get the spatial data for the successful catastrophe administration for surge-influenced ranges. The created approach has helped in distinguishing issues that may enhance the present practices of debacle administration process. The methodology gives an appropriate and fast basic leadership instrument for quick reaction to crises if utilized viable, which thusly would help in minimizing death toll and property. Al-Sabhan et al. (2003) proposed a GIS-construct study in light of the improvement of surge demonstrating and representation for Allahabad Sadar Sub-District (India) be introduced. This incorporated the configuration, analyzed the present status of ongoing hydrological models utilized for surge now throwing and danger alleviation, and showed how electronic frameworks can conquer a portion of the impediments of existing frameworks. While hydrological imaginative and vigorous models are accessible, they are ineffectively suited to the constant application and are regularly not very much incorporated with spatial datasets, for example, GIS. The researched framework may need adaptability, adaptability, and openness by a scope of end clients; in any case, it gives frontline answer for compelling basic leadership. The framework takes into consideration intelligent and aggregate preparing of continuous precipitation information from a remote observing system. A spatially appropriated GIS-based model is incorporated by this approaching information, approximating ongoing to create information on catchment hydrology and overflow.

Buchele et al. (2006) presented a modern approach for integrated flood risk assessment. Figure 3 provides a view on flood modeling and simulation technologies. In light of the setting of a more relative examination of various surge hazard evaluation models, for peril and hazard mapping, amid amazing circumstances. The adequacy of synchronous and organized departure procedures utilizing was examined by Chen and Zhan (2008) using agent-based simulation. The study utilized an operator-based strategy to model movement streams at the level of individual vehicles and to research the aggregate practices of emptying vehicles. Reenactment results show that no clearing technique can be considered as the best system crosswise over various street system structures, and the execution of the procedures relies on upon both street system structure and populace thickness. It

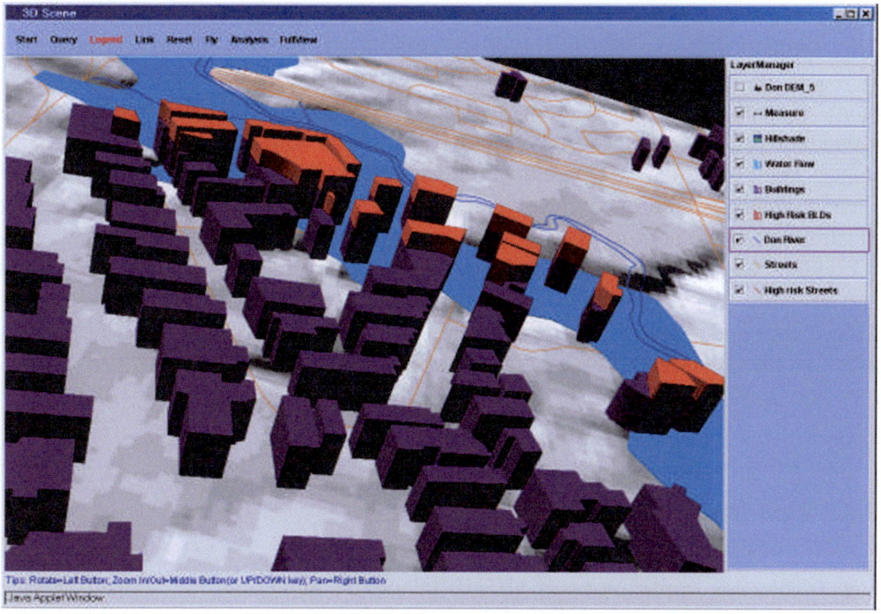


**Fig. 3** Simulation of flood impact on urban center, after (Abdalla and Niall 2009)

has additionally highlighted that the populace thickness in the influenced range is high to be transported on the whole by the street system. De Silva (2000) introduced a model spatial choice emotionally supportive network (SDSS) which was expected for plausibility making courses of action for calamity departure which joins reaction operations with spatial data taking care of and representation limits of a GIS. It interfaces together the geospatial part with the spatial examination segment given by the GIS, with a modernization model proposed to mirror the stream of a clearing procedure in the inconspicuous component. The point has been to plot an SDSS, so that gives an instinctive clearing recreation framework with a component outline that mulls over experimentation with courses of action by giving speedy departure from the reenactment. The thought is that calamity administrators will have the ability to use the SDSS to attempt diverse things with crisis departure courses of action; to get prepared for different potential outcomes.

Moreri et al. (2008) proposed a way to deal with making an electronic Geographic Information Systems (WebGIS) application, which would bolster individuals living in surge fields, who may at one point be defenseless because of their nearness to the stream and the sufficiency of the flooding. The principle result is an electronic GIS framework that gives suitable data to the applicable power and overall population, quickly and for simplicity of presentation to take into account point-by-point understanding. Experienced powers, for this situation, may incorporate surge administrators, surge cautioning specialists, proficient accomplices, and crisis administrations. The simple way to utilize an interface that was produced permit non-GIS specialists, also to be in a position to intelligently see and investigate, and also inquiry the database, to choose distinctive information variables and to see maps at a few levels of the subtle element. Zerger and Wealands (2004) demonstrated that spatially fast hydrodynamic surge models can expect an imperative part in typical peril hazard lessening. A key part of these models that make them appropriate for danger demonstrating is the ability to give time game plan inundation information about the onset, length, and setting off to a crisis circumstance. Such information can be the premise for area use arranging, for mapping, for clearing steering, and for finding reasonable emergency administration to give a few illustrations risk reactions. The examination reasons that powerful calamity hazard decrease is in giving catastrophe supervisors access to model results in a sorted out and versatile structure that licenses results of different danger circumstances to be overviewed and mapped. To address these restrictions, a structure has been made that interfaces business financial databases administration structure with a GIS-based choice emotionally supportive network.

## 2.6 Pandemic Outbreaks

One of the key applications of GIS in pandemic modeling and simulation is to facilitate access to health services by residents who live in and around the security area of a mass gathering event. This will be done by designing an application GIS to assist healthcare authorities in the planning and implementation of the emergency medical response with a focus on optimizing service to vulnerable populations, including (Sharma et al. 2008):

1. Ensuring uninterrupted routine healthcare services during periods of limited access to a security area;
2. Ensuring evacuation procedures for medical emergencies that are nonevent related;
3. Ensuring timely evacuation and health care in the event of a mass casualty incident.

This can be achieved by designing a mapping tool to identify vulnerable populations within the impacted area in case of a pending natural or technological disaster such as a heat wave or power outage (Becerra-Fernandez et al. 2008). GIS will identify access and evacuation routes for impending or in-progress emergency or disaster events. Destinations may include shelters, schools, or other predefined sites outside of the security zone (Chandana et al. 2007). The key support of GIS in a pandemic outbreak can be through providing an appreciation of the use of GIS in addressing public health issues, specifically, to define its uses and limitations in dealing with the questions of defining vulnerable populations. GIS supports possible interventions such as (Daley et al. 2015):

1. Choosing sites for community influenza clinics and vaccination stations
2. Monitoring and evaluating impact of immunization clinics and stations
3. Canceling public events, meetings, and gatherings
4. Closing schools, public places, and office buildings
5. Restricting use of public transportation systems
6. Identifying potential group quarantine and isolation facilities
7. Enforcing community or personal quarantines.

## 2.7 *Extreme Heat Attacks*

Disasters and health impacts, in particular on vulnerable populations; and access to health care; GIS are usable in extreme heat attacks in the following (Cioccio and Michael 2007).

**GIS applicability**

The scope and limitations of the GIS literature search relate to its applicability in health situations, and was targeted but not restricted to mass gatherings, vulnerable populations, and disasters. The intent was to survey the many applications of GIS in health through review of abstracts and conduct a more thorough review of methodology and practical applications to the three domains listed above in articles selected.

## 2.8 Mass Gathering and Civil Unrest

Many types of mass gatherings and the populations attend them vary accordingly. For example, civil demonstrations, outdoor rock concerts, and a football match. These events normally do not attract one kind of attendee, and risks may be associated with weather-related illness, toxic effects of drugs, or trauma due to attendees trying to get close to the stage (McDonald 2008). Political events such as political party conventions or summits have different risks, which include trauma or toxic effects of dispersion agents during political protests, and terrorism-related incidents (Bradler et al. 2008). GIS-related literature revealed that GIS can be effective in Becerra-Fernandez et al. (2008):

1. Specifying the distribution of people around the event area.
2. Analyzing the scope and approach for mapping evacuation in case of emergency.
3. Determining the positions and direction of a move for law enforcement in the field.
4. Analyzing the pattern of movement of masses.
5. Supporting effective decision-making regarding evacuation and response to an emergency.

## 2.9 Sandstorms

Dust storms are also known as sandstorms represent one of the natural hazards with a broad range of environmental impacts; the occurrence of a dust storm impact human health in different ways. Dust storms are a significant cause of traffic accidents and cause air transportation delays. The eruption of a storm introduces fine particles, salts, and chemicals (including herbicides) into the atmosphere, with a suite of health impacts, including not only respiratory complaints but also other serious illnesses. Dust storms can transport allergens including bacteria and fungi, thus impact human health (Goudie 2008).

The recent developments in global warming and climate change have led to increased activity of dust storms in different parts of the world. Many scholars including Goudie (2008), Xu et al. (2006) have worked on the investigation of sand–dust storm events and land surface characteristic using GIS and Remote Sensing. The main procedure depends on the analysis of weather stations data and visualization of the spread of particulate matter in certain space in connection to Dry Mid Temperature and Sub-Dry Temperature, specifically in the desert or semidesert zones (Goudie 2009). Statistical analyses demonstrate that the occurrence of sand–dust storms associate extremely with wind speed, which in turn is strongly related to land surface features; on the other hand, a significant correlation between storm events and other atmospheric quantities such as precipitation and temperature were not observed. This is in addition to the factor of vegetation cover, which has been strongly correlated to dust storms.

Spatial Analysis can be very effective in modeling and visualizing the extent and the impact of dust storms. In particular, we can use GIS to provide the following capabilities in dealing with dust storms emergency management.

## 3 Challenges and Future Directions

Numerous occasions, including the Indian Ocean wave of 2005, the Hurricanes of the 2005 season, and the 7/7 and 9/11 terrorist assaults, have made every one of us intensely mindful of the weakness of the modern society (Goodchild 2006). Knowing the historical record of the events and where such events have occurred, in addition to the geographic limits of their impacts are evidently necessary, principally when combined with information on human populations, infrastructure, and other spatially distributed phenomena that may be relevant to response and recovery (Abdalla et al. 2014). In any case, GIS and spatial technologies that collect, analyze, and allow for visualization of such data (GIS, remote detecting, GPS, and Photogrammetry.) are unmistakably vital in all parts of the disaster management cycle, from protection, response, and recovery through recognition to the reaction and possible recuperation. Geographic Information and Technology (GI&T) gives the premise to assessing and mapping hazard, for arranging evacuation routes and shelters. It also allowed on deciding ranges where human populations are well on the way to have been affected by a disaster, and for assigning resources during the recovery process, among numerous other indispensable and vital tasks that the GIS brings (Abdalla 2015).

One of the observed challenges in the application of GIS for urban emergency management is the location dependency of the event, or what is known as the geographic interdependence of the event. The proximity of the event can lead to complexity in determining the space and location coverage of the case because the occurrence of multiple incidents in the same location can result in cascading or escalating effects among the differently impacted entities.

Regarding the utility of Spatial Analysis in coordination policies, the status of the spatial analysis faces significant deficiencies that can be summarized by the lack of some standard procedures in some regions as well as the unstructured protocols that are in use when dealing with complex scenarios of emergencies in different parts of the world. Part of this can also be attributed to the lack of training in using GIS systems, which can sometimes cause first responders to act inefficiently when dealing with disasters. Spatial Analysis can support effective training for emergency management personnel to deal with complex situations. More importantly, the justified need for geospatial data in some regions of the world requires effective access to global SDI using advanced access protocols and to support the development of procedures and methods for decision-making. This can be attributed to the lack of interoperability in data exchange and standardization.

The decision-making processes face some challenges regarding the lack of common operating picture for the decision-making process. Decision-makers require to access spatial models effectively in the form of real-time data feeds from planning groups, from field observers and remotely collected data. This supports the effective access to operations fields monitoring devices and also the effective decision-making process.

## 4 Conclusions

The spatial analysis provides effective means in dealing with hazard and risk mapping and assessment. It is also effective in providing visual models that help decision-makers to utilize these technologies effectively. However, spatial analysis cannot be effective without accessing accurate data, formulating efficient policies, and securing effective performance of human resources involved in Emergency Management.

The scope of this research is challenging, with an attempt to cover several environmental hazards that might cause disasters in urban regions. It is imperative to deal effectively with emergency management in any of its four phases, i.e., preparedness, mitigation, response, and recovery is crucial. The detailed discussion of the state-of-the-art spatial analysis in urban emergency management has raised some issues related to the maturity of the use of spatial technologies in emergency management, particular issues are:

1. The co-locality of impact by multiple events might require more advanced spatial analysis solutions for providing information about the event location and the extent of the impact.
2. Issues with data and systems interoperability are crucial in providing a timely solution through spatial analysis.
3. Health-related emergencies are more complex to analyze spatially due to issues related to private access to patient's information, as well as the difficulty of covering multiple scales of events.
4. Although the current GIS systems provide advanced capabilities of spatial analysis, yet new approaches for analysis, visualization, and integration are required to provide additional means of support to urban emergency management community.

## References

Abbas SH, Srivastava RK, Tiwari RP, Bala Ramudu P (2009) GIS-based disaster management: a case study for Allahabad Sadar subdistrict (India). Int J Manag Environ Qual 20(1):33–51

Abdalla R (2015) Strategic framework for advancing the utility of GeoICT in emergency preparedness. In: 2nd international conference on the use of ICT for disaster management, Rene, France, IExplorer

Abdalla R, Niall KK (2009) WebGIS-based flood emergency management scenario. The international conference on advanced geographic information systems & web services (GEOWS). IEEE Computer Society Press

Abdalla R (2016) Evaluation of spatial analysis for urban emergency management. SpringerPlus 5:2081. doi:10.1186/s40064-016-3723-y

Abdalla R, Elawad Y, Chen Z, Han SS, Xia R (2014) A GIS-supported fuzzy-set approach for flood risk assessment. Can Water Resour J (Revue canadienne des ressources hydriques) 39 (1):3–14

Al-Sabhan W, Mulligan M, Blackburn GA (2003) A real-time hydrological model for flood prediction using GIS and the WWW. Comput Environ Urban Syst 27:9–32

Anjum M, Rana M, Ivanova J, Abdalla R (2011) GIS-based emergency management scenario for urban petroleum storage tanks. 2011 international workshop on multi-platform/multi-sensor remote sensing and mapping, p 5

Bally P, Bequignon J, Arino O, Briggs S (2005) Remote sensing and humanitarian aid—a life-saving combination. Bulletin of the European Space Agency (ESA), pp 37–41

Becerra-Fernández I, Prietula M, Valerdi R, Madey G, Rodríguez D, Wright T (2008) Design and development of a virtual emergency operations center for disaster management research, training, and discovery. In: Proceedings of the 41st annual Hawaii international conference on system sciences

Bradler D, Kangasharju J, Mühlhäuser M (2008) Systematic first response use case evaluation. In: Second international conference on pervasive computing technologies for healthcare, 2008

Büchele B, Kreibich H, Kron A, Thieken A, Ihringer J, Oberle P, Merz B, Nestmann F (2006) Flood-risk mapping: contributions towards an enhanced assessment of extreme events and associated risks. Nat Hazards Earth Syst Sci 6:485–503

Camps R (1993) Controlling the danger zone. New Sci 138(4):16–17

Chandana S, Leung H, Levy J (2007) Disaster management model based on modified fuzzy cognitive maps. In: IEEE international conference on systems, man and cybernetics, 2007. ISIC

Chen X, Zhan FB (2008) Agent-based modelling and simulation of urban evacuation: elative effectiveness of simultaneous and staged evacuation strategies. J Oper Res Soc 59:25–33

Cioccio L, Michael EJ (2007) Hazard or disaster: tourism management for the inevitable in Northeast Victoria. Tour Manag 28(1):1–11

Correia FN, Rego FC, Saraiva MDG, Ramos J (1998) Coupling GIS with hydrologic and hydraulic flood modelling. Water Resour Manag 12:229–249

Cova TJ (1999) GIS in emergency management. In: Longley MFGPA, Maguire DJ, Rhind DW (eds) Geographical information systems: principles, techniques, applications, and management. Wiley, New York, pp 845–858

Cova TJ, Church RL (1997) Modelling community evacuation vulnerability using GIS. Int J Geogr Inf Sci 11(8):163–784

Currion P (2006) Emergency capacity building project information technology and requirements. Assessment report global. 2006, emergency capacity building project. I. W. Group, London, p 34

Daley R, Ferreras F, Orly J, Abdalla R (2015) GIS for pandemic zoning: application of Brampton, Ontario, Canada. Comput Sci Inf Technol 3(5):149–158

De Silva FN (2000) Integrating simulation modelling and GIS spatial support systems for evacuation planning. J Oper Res Soc 51(4):423–430

ESRI (1999) GIS for emergency management. ESRI White Paper

Eveleigh TJ, Mazzuchi TA, Sarkani S (2007) Spatially-aware systems engineering design modeling applied to natural hazard vulnerability assessment. Syst Eng 10(3):16–24

Goodchild MF (2006) GIS and disasters: planning for catastrophe. Comput Environ Urban Syst 30 (2006):227–229

Goudie AS (2008a) Dust storms: recent developments. J Environ Manag 90:89–94

Goudie AS (2008b) The history and nature of wind erosion in deserts. Ann Rev Earth Planet Sci 36:97–119

Goudie AS (2009) Global deserts and their geomorphological diversity. In: Parsons AJ, Abrahams AD (eds) Geomorphology of desert environments, 2nd edn. Springer, London, pp 9–20

Jaiswal RK, Mukherjee S, Raju KD, Saxena R (2002) Forest fire risk zone mapping from satellite imagery and GIS. Int J Appl Earth Obs Geoinf 4(1):1–10

Johnson K (2003) GIS emergency management for the University of Redlands. In: ESRI international user conference

Kumar A (2013) Natural hazards of the Arabian Peninsula: their causes and possible remediation. Earth system processes and disaster management. Society of Earth Scientists Series, pp 155–180

Kwan M, Lee J (2005) Emergency response after 9/11: the potential of real-time 3D GIS for quick emergency response in micro-spatial environments. Comput Environ Urban Syst 29(2):93–113

McDonald R (2008) New considerations for security compliance, reliability and business continuity. In: 2008 IEEE rural electric power conference

Miura H, Yamazaki F, Matsuoka M (2007) Identification of damaged areas due to the 2006 Central Java, Indonesia earthquake using satellite optical images. In: Urban remote sensing joint event, 2007

Moreri KK, Mioc D, Anton F, Nickerson B, McGillivray E, Morton A, Ulieru M, Tang P (2008) Web based geographic information systems for a flood emergency evacuation. In: 3rd international ISCRAM China workshop, Harbin, China

Morrow B (1999) Identifying and mapping community vulnerability. Disasters 23(1):1–18

Pew KL, Larsen CPS (2001) GIS analysis of spatial and temporal patterns of human-caused wildfires in the temperate rain forest of Vancouver Island, Canada. Forest Ecol Manag 140(1):1–18

Pradhan B, Suliman MDHB, Awang MAB (2007) Forest fire susceptibility and risk mapping using remote sensing and geographical information systems (GIS). Dis Prev Manag 16(3):344–356

Saadatseresht M, Mansourian A, Taleai M (2009) Evacuation planning using multi objective evolutionary optimization approach. Eur J Oper Res 198(2009):305–314

Sharma R, Kumar SB, Desai NM, Gujraty VR (2008) SAR for disaster management. IEEE Aerosp Electron Syst Mag 23(6):4–9

Smirnov A, Pashkin M, Chilov N, Levashova T (2006) Context-based disaster management support. DIS 2006: IEEE workshop on distributed intelligent systems: collective intelligence and its applications, 2006, pp 291–296

Verjee F (2005) The Application of geomatics in complex humanitarian emergencies. J Humanitarian Assistance: 7–8

Xu X, Levy JK, Zhaohui L, Hong C (2006) An investigation of sand–dust storm events and land surface characteristics in China using NOAA NDVI data. Glob Planet Change 52:182–196

Zerger A, Wealands S (2004) Beyond modelling: linking models with GIS for flood risk management. Nat Hazards 33:191–208

## Author Biography

**Dr. Rifaat Abdalla** is Associate Professor and Head, Department of Hydrographic Surveying, Faculty of Maritime Studies, King Abdulaziz University, Jeddah, Saudi Arabia. His current research interests are focusing in GIS and Web-based GIS for Disaster Management, and

Visualization for Disaster and Emergency Management Decision Support. Dr. Abdalla has published in reputed international journals in the area of GeoICT and Disaster Management. Dr. Abdalla has received numerous international awards including the 2008 Second Place ESRI Best Scientific Paper in GIS, Awarded by the American Society for Photogrammetry and Remote Sensing. Also, he received GeoWeb Conference Best Paper in 2009 and Natural Science and Engineering Research Council of Canada Doctoral Scholarship and Postdoctoral Fellowship. Dr. Abdalla is a member of Ontario Association of Professional Geoscientists and a Certified GIS Professional, by the American GIS Certification Institute.

# Experimental Study of the Mechanics of Gypsum Seam Hazard for Abu Dhabi

**M. Opolot, W. Li, R.L. Sousa and A.L. Costa**

**Abstract** Abu Dhabi is predicting a huge growth in population over the next 20 years (Plan Abu Dhabi 2030); further, it seeks to become an international destination for tourists, businesses, and investment while protecting its cultural heritage. A crucial aspect of achieving this goal is the development of large integrated transportation system, underground, and above ground, to ensure Abu Dhabi becomes a sustainable city on a global scale. The presence of gypsum rocks that occurs within Abu Dhabi's bedrock is a major threat to underground construction and understanding the phenomena is of paramount importance. They are persistent quasi-horizontal bands, at different levels (top level between 10 and 15 m and bottom level between 15 and 25 m), prone to volume change by dissolution or swelling, due to changes in the stress regime and water chemistry and flow. The dissolution of gypsum is also a cause for cavities that can be found within this formation in greater Abu Dhabi. In this paper, the description and results of an experimental study aimed at obtaining a better understanding of the gypsum dissolution process, as well determine factors affecting it, are presented. Tests on the dissolution process of gypsum rock were performed using artificially created intact and fractured gypsum samples which are a representative of the collected in situ fractured gypsum rock samples obtained from Abu Dhabi. The samples are subjected to flow-through tests. Results obtained show that for an initially saturated gypsum specimen, there is a sharp decline in concentration with time (Stage I), followed by a constant concentration (Stage II) before a slight gradual increase is

M. Opolot (✉) · R.L. Sousa · A.L. Costa
Masdar Institute of Science and Technology, Abu Dhabi, UAE
e-mail: mopolot@masdar.ac.ae

R.L. Sousa
e-mail: rsousa@masdar.ac.ae

A.L. Costa
e-mail: acosta@masdar.ac.ae

W. Li
Massachusetts Institute of Technology, Cambridge, USA
e-mail: weili08@mit.edu

 
S. Pirasteh and J. Li (eds.), *Global Changes and Natural Disaster Management: Geo-information Technologies*, DOI 10.1007/978-3-319-51844-2_16

observed (Stage III) with time. This is a fundamental study—part of a larger set of experiments studying the gypsum dissolution process in Abu Dhabi. Using the data collected from the field and the experiments mentioned above, gypsum geohazard risk-related maps which reflect subsidence, swelling, cavity collapse, and cavity flooding associated with Gypsum Karst shall be developed using Geographic Information Systems.

**Keywords** Gypsum dissolution · Groundwater flow · Underground construction · Subsidence · Collapse

## 1 Introduction

Gypsum ($CaSO_4.2H_2O$)—a white, translucent, pale grey or grey rock mineral—is known to occur worldwide in all geological sequences from the Cambrian (570 million years ago) to the recent deposits (Cooper and Calow 1998). In Abu Dhabi, gypsum occurs at different levels (top level between 10 and 15 m and bottom level between 15 and 25 m) within the Tertiary bedrock as persistent quasi-horizontal bands. Figure 1 shows samples from representative gypsum layers, retrieved at different depths in Abu Dhabi.

**Fig. 1** Abu Dhabi gypsum rock samples

Gypsum is soluble in water, and it dissolves about 100 times faster than limestone (Yilmaz et al. 2011). In pure water at a temperature of about 200 °C, the solubility of gypsum is 2.531 g/L (Klimchouk 1996). Water flow velocity, temperature, salinity, applied pressure, gypsum content percentage, gypsum grain size, time of groundwater exposure are among other factors that influence the dissolution rate of gypsum. Gypsum is further affected by diagenesis—a process where primary (depositional) gypsum is transformed to anhydrite ($CaSO_4$) when buried. Anhydrite can also rehydrate to form gypsum if it is exposed to a weathering process under favourable conditions. Gypsum is also prone to volume change by dissolution or swelling due to stress regimes and groundwater chemistry as shown in Eq. (1) (Yilmaz et al. 2011; Llana-Fúnez et al. 2012).

$$\underset{\text{gypsum}}{CaSO_4 \cdot 2H_2O} \rightarrow \underset{\substack{\text{bassanite} \\ \text{total volume increase} \rightarrow \\ \leftarrow \text{solid volume increase}}}{CaSO_4 \cdot \frac{1}{2}H_2O} + \underset{\text{water}}{\frac{3}{2}H_2O} \tag{1}$$

When gypsum is exposed to groundwater flow, it may dissolve with the development of karst features. Continuous exposure to groundwater can further lead to subsidence or even collapse of the overlying ground surface. This is a major threat for underground construction over gypsiferous terrains—for which Abu Dhabi is part of—and understanding the phenomena is of paramount importance.

## 2 Abu Dhabi Geological Setting

### *2.1 Geology and Geotechnical Description*

The Emirate of Abu Dhabi (see Fig. 2) alone is approximately 80% a sandy desert, and it covers about 85% of the entire United Arab Emirates land area (Shabbir et al. 2008). The Emirate is located in the north-east portion of the Arabian Plate, that formed during the Late Neoproterozoic era ($\sim$820–750 Ma) by the accumulation of island arcs and microcontinents to early Gondwana (Glennie 2013). The Zagros foreland basin and adjacent fold-belt bound Abu Dhabi to the north, the Omani thrust belt to the east, the Rub al Khali Basin to the south, and Qatar Arch to the west (Salah 1996).

In the present day Abu Dhabi, desert land surface sits on rocks deposited nearly 950 million years ago. The majority of the depositions happened below sea level. However, during lengthy periods of land surface uplift—an average rate of 3 mm/year for the past 25,000 years (Lokier and Steuber 2008)—and subaerial erosion, geological traces over that time period were washed away. The geological history of that time period is buried underneath the land surface and the Arabian Gulf (Shabbir et al. 2008), composed of various stratigraphic layering, including

**Fig. 2** Reference map of the Emirate of Abu Dhabi (Al-Katheeri 2008)

gypsum. A stratigraphic geological profile drawn during the fieldwork activity at about 20 km away from Masdar City (see Fig. 3) shows that there is an occurrence of gypsum at just about 30 m. A comparison with the Masdar City construction geotechnical reports shows bands of gypsum at almost the same depth. This, therefore, implies that areas with the same geological formation tend to have the same layering, hence in Abu Dhabi, gypsum occurrence can be noticed at just tens of metres below sea level.

## 2.2 *Hydrogeology of Abu Dhabi*

The Emirate of Abu Dhabi experiences arid desert conditions with minimal amounts of annual rainfall ranging from 0 to 30 mm/annum (Brook and Houqani 2003). It is also characterized by the absence of vegetation and very saline groundwater conditions. In addition, groundwater recharge is less than 4% of total annual rainfall, and there are no potential perennial surface water resources (Brook and Houqani 2003; Al-Katheeri 2008). In Abu Dhabi, groundwater lies in consolidated or unconsolidated surficial aquifers and as bedrock/structural aquifers (Shabbir et al. 2008).

Gypsum dissolution is enhanced by groundwater flow conditions. Groundwater flow systems can be divided into local, intermediate and regional flow systems. Flow nets that predict groundwater flow can be derived based on each of these

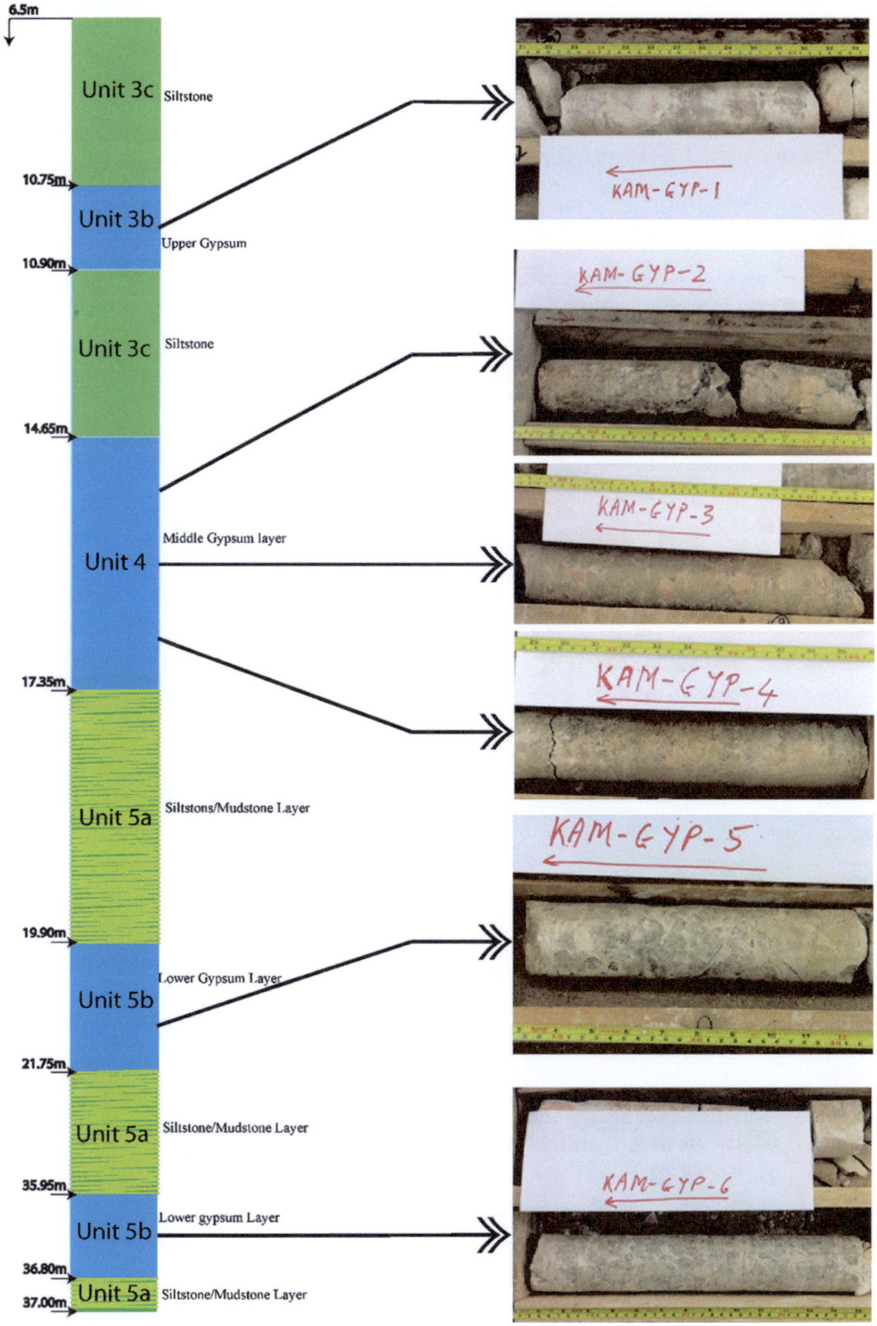


**Fig. 3** Stratigraphy of the borehole located at 20 km away from Masdar City

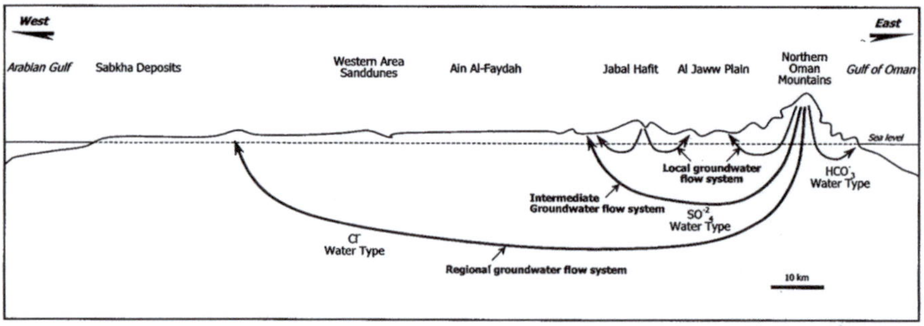


**Fig. 4** Schematic showing groundwater flows regime in Abu Dhabi (Alsharhan et al. 2001)

**Fig. 5** Karst features within Abu Dhabi (Modified) (Tayo 2014)

scales, and are dependent on many factors that include local topography and basin shape geometry. Based on the study by Toth (1963) and Alsharhan et al. (2001) came up with the schematic illustrating a flow regime in Abu Dhabi (see Fig. 4).

## 3 Gypsum Seam Hazards

Several cases of destruction by gypsum rock dissolution, subsidence and collapse have been recorded worldwide dating back to the nineteenth century. The gradual destruction process of gypsum rock dissolution can cause failure of tunnels and other infrastructures even after decades of operation (Kaiser et al. 2010). In-depth records of gypsum hazards are highlighted in the studies by Cooper (1998, 2006, 2008), Gutierrez and Cooper (2002), Kaiser et al. (2010), and Butscher et al. (2011).

Although there have been no catastrophic gypsum seam hazards recorded in Abu Dhabi yet, this does not prevent its future occurrence. Sinkholes can be observed in some areas of Abu Dhabi (see Fig. 5). This, therefore, calls for an understanding of the mechanics of gypsum seams in Abu Dhabi, which will later support the development of hazards maps for the region. These hazard maps will be developed using GIS based on the experimental data obtained in this study.

# 4 Experimental Setup

## *4.1 Overview*

A mathematical model was developed by Li et al. (2015) to predict the dissolution process of a solution rock based on advection, diffusion and dispersion. The model is capable of predicting the concentration–time curve of the effluent. To validate the mathematical model, Li et al. (2015) and Opolot (2015) performed experiments.

## *4.2 Plaster of Paris Calibration Tests*

To obtain the relationship between electric conductivity and concentration of the Plaster of Paris to be used for gypsum specimen moulding, a batch of calibration tests were performed. This was performed by preparing a saturated solution of Plaster of Paris and its saturated conductivity measured before subsequence dilution using distilled water is performed while measuring the corresponding conductivity up to when the conductivity is almost equal to that of distilled water. Using curve fitting, a log–log plot of the raw data (see Fig. 6) and an equation relating conductivity to concentration is derived as shown in Eq. (2).

$$\log(\text{Conc.}) = 1.225\log(\text{E.C.}) - 3.749 \tag{2}$$

## *4.3 Specimen Preparation*

In order to perform the flow-through tests, gypsum specimens were first moulded using the locally purchased Plaster of Paris. When water is mixed with Plaster of Paris at room temperature, gypsum is obtained as shown in Eq. (3).

$$CaSO_4 \cdot \frac{1}{2}H_2O + \frac{3}{2}H_2O \xrightarrow{\text{room temperature}} CaSO_4 \cdot 2H_2O \tag{3}$$

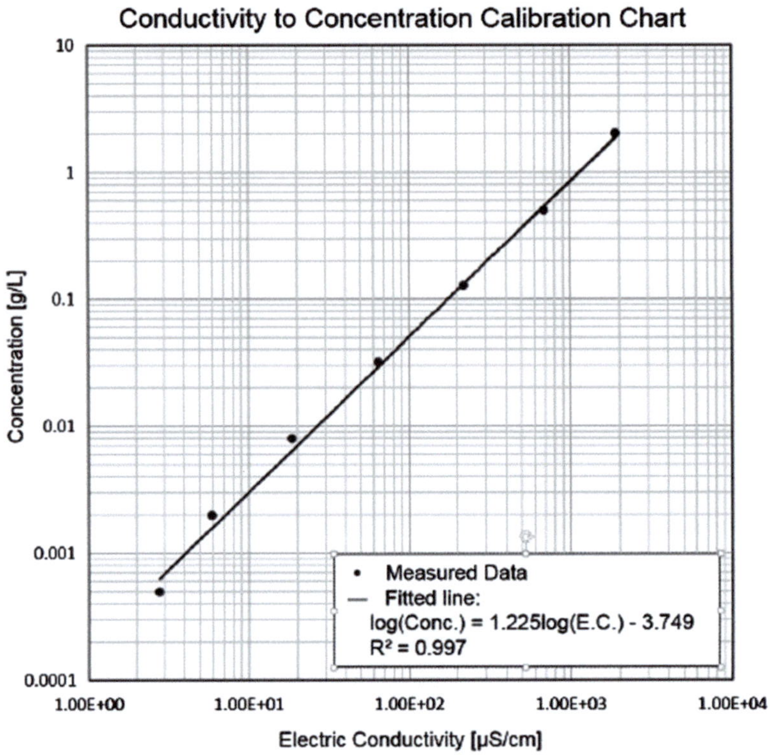


**Fig. 6** Conductivity metre calibration

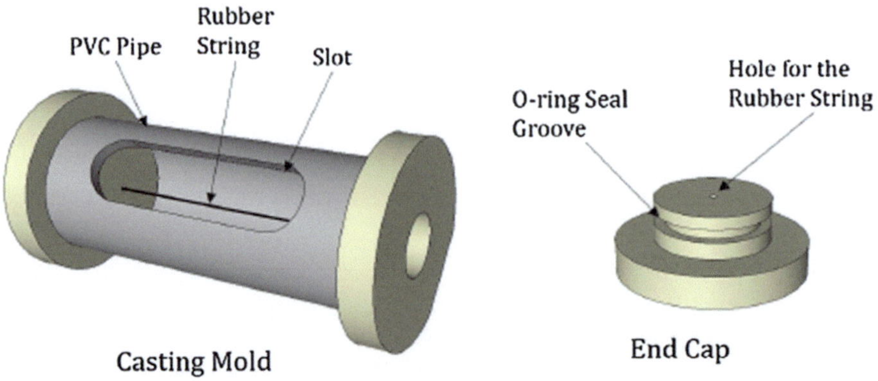


**Fig. 7** Casting mould used for specimen preparation

A ratio of 0.6 of water to Plaster of Paris was poured into the mixture and mixed for about 2 min to obtain gypsum paste. A cylindrical mould (see Fig. 7) with a string of about 1 mm passing through was then placed on the vibrator before the prepared gypsum paste is poured into it to obtain the cylindrical specimen with an initial flow-through hole. The entire mould process is performed on the vibrator to eliminate any trapped air bubbles within the gypsum paste. After moulding, the specimen in transferred to an oven set at 40 °C to cure. In about 24 h, the rubber string is pulled off to leave an initial flow-through hole while the specimen is further cured in an oven. The whole specimen preparation process takes about 7 days to get it ready for flow-through tests. Before the specimen is used for flow-through tests, it first saturated in a fully saturated solution of gypsum. This is done for the specimen to get fully saturated and hence to prevent it from absorbing water during the flow-through batch tests.

## 4.4 Apparatus Setup

The flow-through apparatus consisted of the peristaltic/submerged pump, constant head reservoir, throttle valve and the gypsum specimen assembly as shown in Fig. 8.

The apparatus was first calibrated by letting distilled water flow through it to eliminate any trapped in the air before the specimen is connected at the end. The specimen is held tight using a C-clamp to avoid any leakage during the experimental run. A sampling of the effluent is performed in intervals from the outlet of

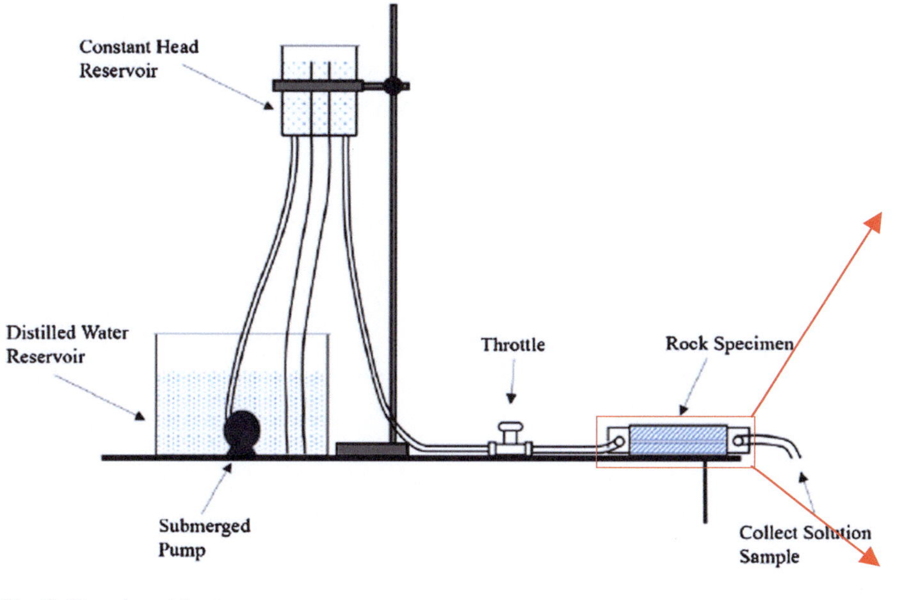


**Fig. 8** Experimental setup

the gypsum specimen. The electric conductivity of the different samples collected is measured and the corresponding concentration calculated using Eq. (2) obtained during calibration.

## 5 Results and Discussions

When the gypsum specimens were fully saturated, flow-through tests were performed by sampling the outflow from the gypsum specimen at a given time interval for analysis. The electric conductivity of each sampled solution was then measured using the conductivity metre and its corresponding concentration calculated using Eq. (2) relating conductivity to concentration. A plot of concentration (g/L) versus time (s) shown in Fig. 9 illustrates that, for an initially saturated gypsum specimen, there is a sharp decline in concentration with time (Stage I), followed by a constant concentration (Stage II) before a slight gradual increase is observed with time (Stage III).

In Stage I, at early times, there is a high initial concentration of dissolved solutes in the water outflow. This concentration is similar in magnitude to the concentration of the dissolved solids in the water inside the specimen that was saturated for 4 days. With time, the injected distilled water dilutes the solution in the gypsum specimen opening, causing a decrease in concentration of the solutes. The time duration of this stage depends on the flow rate of the distilled water, where a lower flow rate leads to a longer dilution duration (Stage I) before Stage II is observed.

In Stage II, an almost constant concentration of dissolved solutes is observed with time. This is because as injected distilled water continues to flow through the gypsum specimen opening, the dissolution processes reach equilibrium, leading to a constant concentration.

**Fig. 9** Concentration variation with time

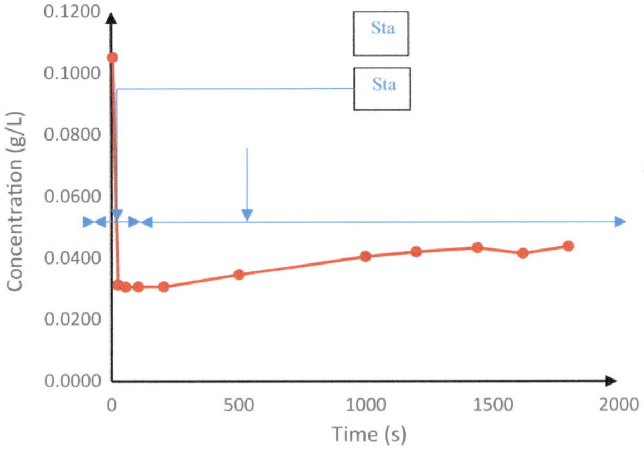

In Stage III, a gradual increase in concentration is observed with time. In this stage, and as water continues to flow through the opening, it dissolves the inner surface wall of the gypsum opening, causing an increase in the opening diameter. The increase in diameter results in a larger contact surface area between water and the gypsum opening. As this happens, the water flow rate decreases, which results in a longer exposure time to the flowing water. The combined effect of these phenomena leads to a longer reactive time between the water and the gypsum, causing a gradual increase in concentration with time.

# 6 Conclusions and Recommendations

## *6.1 Conclusions*

The results of experimental analyses performed on gypsum specimens representative of the naturally occurring gypsum around Greater Abu Dhabi show that gypsum is subject to dissolution once exposed to water. The rate of gypsum dissolution is influenced by the time of exposure of the gypsum specimen to the flowing water.

A constant head experimental setup was used to study the effects of gypsum dissolution. The setup was first calibrated to determine the flow rate of water into the gypsum specimen. Equations relating concentration and conductivity were derived. The calibration charts obtained were in agreement with those suggested by Richards (1954).

The results of the solubility experiments performed in this study were compared to those by Klimchouk (1996). Klimchouk (1996) performed experiments using distilled water at a temperature of about 200 °C, and the solubility of gypsum obtained was 2.531 g/L. The experimental values obtained in this study were about 2.9 g/L for a fully saturated specimen, and hence both results are comparable. Differences in the results may be caused by, amongst other factors, the different experimental conditions.

The results show that at early times of gypsum dissolution, a sharp decline occurs before an almost constant dissolution rate with time is achieved. In general, during the early times, very little gypsum dissolution occurs, and the gypsum dissolution hazard to surface and subsurface infrastructure is low. With time, gypsum dissolution occurs, and the concentration of gypsum in the percolating water gradually increases. As this occurs, surface subsidence and the hazard of gypsum dissolution to surface and subsurface infrastructure may increase. Subsidence and collapse may therefore occur over the course of time.

These results are similar to observations made by Thompson et al. (1998), who suggest that collapse may occur either abruptly when the water flow rate is high, or over the course of time when the water flow rate is low.

The ambitious Abu Dhabi 2030 Vision calls for an aggressive development plan that includes surface and subsurface transportation networks and other critical infrastructure systems. Certain areas of Abu Dhabi where gypsum is naturally occurring are susceptible to gypsum dissolution (subsidence and collapse). The results of the experiments performed in this work represent a necessary step in the preparation and validation of gypsum dissolution hazard maps. These hazard maps can then be used by urban planners and policy makers as decision aids for better and safer land use planning.

## 6.2 Recommendations

The work presented in this paper is a necessary first step in understanding gypsum dissolution processes particularly for naturally occurring gypsum in Greater Abu Dhabi. Further sets of experiments are being performed to better understand the dissolution processes. These experiments include but are not limited to:

1. Using the actual Abu Dhabi in situ natural gypsum to perform experiments.
2. Additional experiments with the same conditions (initial and boundary) as performed in this thesis to be able to obtain an experimental sample of results. That is a repetition of the same experiments to obtain several results rather than relying on the results of a single experiment.
3. Additional experiments under different (more field representative of Abu Dhabi) conditions, for example, different initial concentrations.
4. Prolonged exposure of specimens to percolating water at various temperature and saline conditions.
5. The introduction of ions into the percolating water through the gypsum specimen, from which the pairing effect is monitored and analyzed to determine how it enhances or impedes gypsum dissolution.
6. Perform experiments under stress conditions that replicate the in situ stresses gypsum is subjected to (Triaxial).

The ultimate aim is to better understand the gypsum dissolution process to derive gypsum hazard maps for Abu Dhabi. Masdar and Zayed Cities were chosen for the preliminary hazard mapping. For Zayed City, data obtained from the Spatial Data Division of Abu Dhabi municipality was used. On the other hand, data obtained from Sharp (2010) was used for the study of Masdar City. Figure 10 shows the

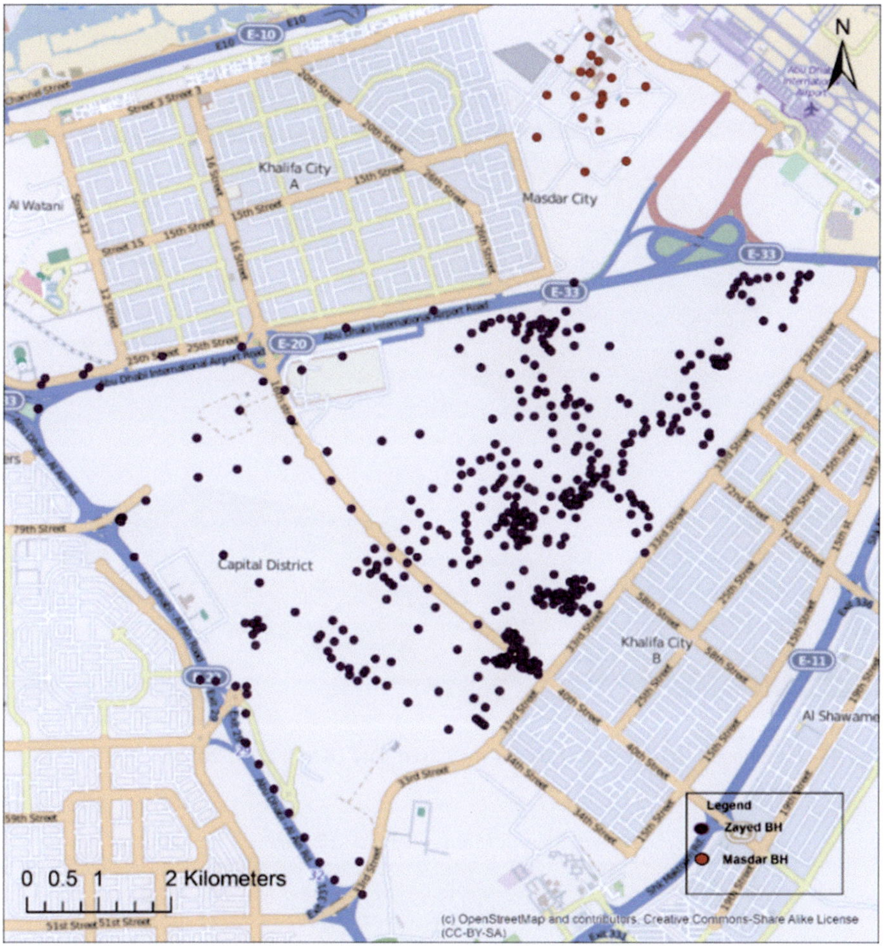


**Fig. 10** Borehole logs for Masdar and Zayed Cities (Kaabi et al. 2015)

boreholes considered for this preliminary study, whereas Figs. 11 and 12 show initial results of how gypsum occurs in the two cities. A full understanding of the sampled cities will lead to the development of a hazard map for the whole Emirate of Abu Dhabi.

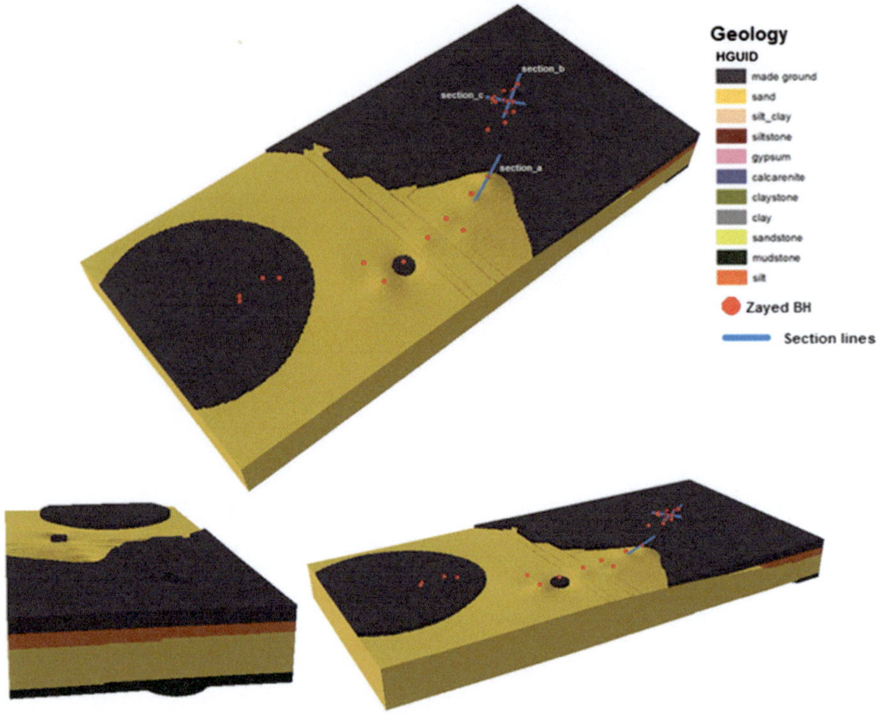


**Fig. 11** Preliminary 3D geological model for Zayed City (Kaabi et al. 2015)

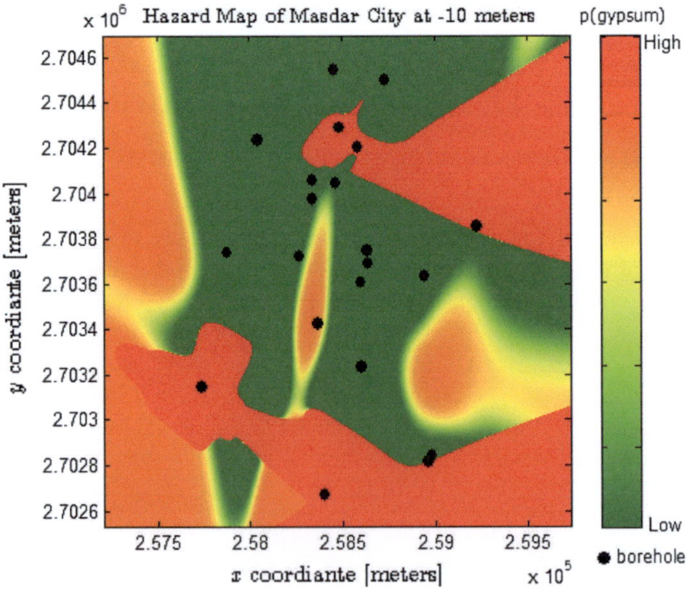


**Fig. 12** Masdar City hazard map at $Z = 10$ m (Abdulla et al. 2015)

**Acknowledgements** This work is part of a one-to-one research project between Masdar Institute of Science and Technology and the Massachusetts Institute of Technology. The authors greatly benefited from the discussions with Dr. Stephen Lokier from the Petroleum Institute, Abu Dhabi. The authors further thank Dr. Lokier for donating natural gypsum rock specimens.

## References

Abdulla MB et al (2015) Subsurface Gypsum Probabilistic Identification Using Artificial Neural Networks. Case Study: Masdar City, United Arab Emirates (Submitted)

Al-Katheeri ES (2008) Towards the establishment of water management in Abu Dhabi Emirate. Water Resour Manag 22:205–215

Alsharhan AS, Rizk ZA, Nairn AEM, Bakhit DW, Alhajari SA (eds) (2001) Hydrogeology of an arid region: the Arabian Gulf and adjoining areas: the Arabian Gulf and adjoining areas. Elsevier, Amsterdam

Brook M, Houqani HA (2003) Assessment of the water situation in the western region of Abu Dhabi Emirate. In: Centre TER (ed), p 1–37

Butscher C, Einstein HH, Huggenberger P (2011) Effects of tunneling on groundwater flow and swelling of clay-sulfate rocks. Water Resour Res 47(11):1–30

Cooper HA (1998) Subsidence hazards caused by the dissolution of Permian gypsum in England: geology, investigation and remediation. Geol Soc London 15:265–275

Cooper HA (2006) Gypsum dissolution geohazards at Ripon, North Yorkshire, UK. The Geological Society of London, pp 1–14

Cooper HA (2008) The classification, recording, databasing and use of information about building damage due to subsidence and landslides. Q J Eng Geol Hydrogeol 41:1–27

Cooper HA, Calow RC (1998) Avoiding gypsum geohazards: guidance for planning and construction. British Geological Survey, p 57

Glennie KW (2013) Structural and stratigraphic evolution of Abu Dhabi in the context of Arabia. In: Lithosphere dynamics and sedimentary basins: the Arabian plate and analogues. Springer, Berlin, pp 3–21

Gutierrez F, Cooper HA (2002) Evaporite dissolution subsidence in the historical city of Calatayud, Spain: Damage appraisal and prevention. Nat Hazards 25:259–288

Kaabi MA et al (2015) Decision support system for tunneling construction: Abu Dhabi case study. In: 3rd Arabian tunneling conference, Dubai (in press)

Kaiser PK, Amann F, Steiner W (2010) How highly stressed brittle rock failure impacts tunnel design. In: Rock mechanics and environmental engineering. Paper presented at European rock mechanics symposium

Klimchouk A (1996) The dissolution and conversion of gypsum and anhydrite. Int J Speleol 25(3):2

Li W, Sousa RL, Einstein HH (2015) Reactive transport in a soluble rock pipe: a theoretical and experimental study. In 2015 AGU fall meeting

Llana-Fúnez S, Wheeler J, Faulkner D (2012) Metamorphic reaction rate controlled by fluid pressure not confining pressure: implications of dehydration experiments with gypsum. Contrib Miner Petrol 164(1):69–79

Lokier S, Steuber T (2008) Quantification of carbonate-ramp sedimentation and progradation rates for the late Holocene Abu Dhabi shoreline. J Sediment Res 78(7):423–431

Opolot M (2015) Experimental study of the mechanics of gypsum seam hazard for Abu Dhabi. In: Mechanical and Materials Engineering Department. Masdar Institute of Science and Technology

Richards LA (1954) Diagnosis and improvement of saline and alkali soils. Soil Sci 78(2):154

Salah M (1996) Geology and hydrocarbon potential of Dalma Island, offshore Abu Dhabi. J Pet Geol 19(2):215–226

Shabbir S et al (2008) Physical geography of Abu Dhabi Emirate United Arab Emirates. Environmental Data Initiative-Environment Agency, pp 1–128

Sharp P (2010) Site-wide infrastructure design geotechnical interpretive report. In: Third Issue. Mott MacDonald

Tayo R (2014) Typical geotechnical risks in the middle east. In: A presentation. Hyder

Thompson A, Hine P, Peach D, Frost L, Brook D (1998) Subsidence hazard assessment as a basis for planning guidance in Ripon. Geol Soc, Lond Eng Geol Special Publ 15(1):415–426

Toth J (1963) A theoretical analysis of groundwater flow in small drainage basins. J Geophys Res 68(16):4795–4812

Yilmaz I, Marschalko M, Bednarik M (2011) Gypsum collapse hazards and importance of hazard mapping. Carbonates Evaporites 26(2):193–209

# Index





S. Pirasteh and J. Li (eds.), *Global Changes and Natural Disaster Management: Geo-information Technologies*, DOI 10.1007/978-3-319-51844-2

# Index





S. Pirasteh and J. Li (eds.), *Global Changes and Natural Disaster Management: Geo-information Technologies*, DOI 10.1007/978-3-319-51844-2

MIX
Papier aus verantwortungsvollen Quellen
Paper from responsible sources
FSC® C105338

If you have any concerns about our products, you can contact us on
**ProductSafety@springernature.com**

In case Publisher is established outside the EU, the EU authorized representative is:
**Springer Nature Customer Service Center GmbH**
**Europaplatz 3, 69115 Heidelberg, Germany**

Printed by Libri Plureos GmbH
in Hamburg, Germany